Contents

Introduction

Science has changed our lives for the better and will continue to do so. The advances of the past one hundred years boggle the mind. Things that were once unimaginable – surround sound, mobile phones, WD-40 – are now indispensable in our daily lives.

Science was the first academic pursuit in our life. As youngsters we saw new things and wondered 'why?' Hopefully, you still wonder why when you see something new. The process of examining the world around us is loosely called science. And science (like all things) is easier to understand when it's mixed with a healthy dollop of humour.

In the last few years, several things have become clear to me. One, learning needs to be fun, whether you are eight or eighty. Almost all public speakers break the ice with a joke before they push their agenda, because you learn better when you're laughing. Two, we need to embrace our manliness. Be glad that men and women are different. For instance, we have a different language from women. We say sweat, she says perspire. Blokes need textbooks written in their language. I love science and being a man, so combining the two seemed like a perfect idea – a science textbook written for the male of the species. So this book is designed to educate and amuse. And the best part is... you won't have to take a final exam when you finish reading it.

ESSENTIAL SH!T
PUB SCIENCE
TO IMPRESS YOUR MATES
Answers all the crazy science questions you might have
D&C
d and Charles

A DAVID & CHARLES BOOK

David & Charles is an F+W Media Inc. company
4700 East Galbraith Road
Cincinnati, OH 45236

First published in the UK in 2011

The material in this book has been previously published in *How Do You Light a Fart?* published by Adams Media, 2009.

A catalogue record for this book is available from the British Library.

ISBN-13: 978-1-4463-0044-2 paperback
ISBN-10: 1-4463-0044-7 paperback

Printed in China by RR Donnelley
for David & Charles
Brunel House, Newton Abbot, Devon

Senior Acquisitions Editor: Freya Dangerfield
Desk Editor: Felicity Barr
Project Editor: Stuart Robertson
Proofreader: John Skermer
Design Manager: Sarah Clark
Production Controller: Bev Richardson

David & Charles publish high quality books on a wide range of subjects.
For more great book ideas visit: www.rubooks.co.uk

Men are different in the humour department also. We often see something to laugh about where most women never would. We smile when our dog slopes off to the other side of the room after polluting our air with noxious fumes. We make jokes about erectile dysfunction medicine (at least until we need it). We smile as we ponder what life was like before the remote control was invented. We laugh about breaking wind, burnt knuckle hair and women.

You will learn about all sorts of manly things in the following pages. This book will teach you the science of flatulence, cool bar tricks and why three-piece balls are better for golf. You will also be taught why space explosions are better in the cinema and at least three ways to open a locked door.

This leads me to make this public service announcement: cherish this book forever and buy copies for all of your law-abiding friends. In your will, leave instructions for it to be given to a charity. But NEVER donate it to any books-for-prisoners charity. Also never give this book as a breakup present to your soon-to-be ex-girlfriend.

Now that you've been forewarned, sit down and hang on as we examine the everyday science of being a bloke.

1. Things that go boom

Men love explosions and our collective fascination with fire and explosions is something inherently male. Hollywood has recognized and catered to that male lust for years with a plethora of action films every summer. In the real world cars almost never explode on impact, but every stunt car crash is accompanied by a raging fireball that stretches a thousand feet into the air. The film industry spends millions of dollars to get bums on seats with tons of high-tech pyrotechnics because they know the secret to cinematic success: men will gladly fork out to watch massive explosions. The movie can be completely crap, but most blokes will still leave the cinema happy if there have been a decent amount of jaw-dropping explosions.

Grill manufacturers have also learned this secret to success. One of the reasons men are addicted to cooking on a barbecue is because of the fun – and danger – that comes from lighting it. Throwing a match onto a pile of lighter fluid-laden charcoal briquettes is the highlight of many picnics. Most blokes would never wax any part of their body, but we'll gladly risk burning off our facial hair for the thrill of that momentarily out-of-control flame. The wonderful whoosh sound and the immediate six-foot-high wall of flame have signified the beginning of many glorious testosterone-filled hours spent standing happily over the grill. The chance that we may lose our eyebrows is only an added bonus. So light a fire, say goodbye to your body hair and put on your flameproof suit as we examine the world of pyrotechnics.

Scientifically speaking

Men love to wonder and that is the nature of science

- Ralph Waldo Emerson

How is an aerial firework display made?

Gigantic firework displays explode in all manners of colours, drawing cheers from anyone watching. It will often take days for pyrotechnical experts to set up for that twenty-minute display. Let's peek inside an individual firework to see how it works.

The outside part of a firework shell is usually formed from paper or thin cardboard. Running down the centre of the shell is a large firecracker-like bursting charge. Surrounding the centre is black powder, often mixed in with pieces of aluminium, steel, iron, zinc or magnesium that create shiny sparks as they burn. The shells also contain stars – sparkler-like devices that account for the falling light trails we see. Different chemicals are added to the stars to create the colours as they burn. You may have learned (or at least were probably taught) the secret to colours in a chemistry lesson; all elements burn with a characteristic flame colour. Common elements used in fireworks are strontium for red, barium for green, copper for blue, magnesium for white and sodium for yellow. The way the stars are packed in the firework shell accounts for the different shapes when the fireworks explode. Ovals, circles and palm trees are all created by positioning the stars certain distances away from the bursting charge. Some shells also have smaller individual shells inside that explode as they fall to create even more patterns.

Big fireworks are fired out of a mortar by the use of a lifting charge. The lifting charge shoots the firework into the air and also ignites a fuse that explodes the bursting charge at a certain height. A show may have hundreds of different mortars all set to be launched in a particular order. Display packs for use in your back garden at home can also be bought. The shells are smaller, but the science is the same.

What is the difference between dynamite and TNT?

Contrary to popular belief, dynamite and TNT are not the same thing. Whereas dynamite is nitroglycerine-soaked sawdust, TNT is a specific chemical compound. TNT is much easier to spell than trinitrotoluene (and dynamite), so it is perfect for cartoons and action movies. It is a yellow crystal that was originally used as a yellow dye, but once men realized the dye tended to explode they found uses that were a little more fun. TNT is extremely stable and very hard to detonate, but when it does, any men standing around cheer. Or crap their pants, depending on how close they are.

TNT can be bought in blocks by itself, but it is more commonly combined with other materials to form an even stronger explosive sold only to professionals. This is a good thing because, given the chance, many men would buy it simply to liven up a Guy Fawkes' night bash.

Did you know?

TNT was originally used as a yellow clothing dye.

How does a large building implode into a neat pile of rubble?

Bringing down a twenty-storey building takes the utmost care, a ton of scientific know-how and a few hundred kilos of explosives. Building implosions are small, controlled explosions designed to bring a building down on its own footprint. Blasters spend years perfecting their craft and most implosions worldwide are handled by the same few companies. But let's face it; these blasters probably grew up blowing stuff up in their dads' sheds and now it's their full-time job, only the explosives and the sheds are a lot larger.

Often blasters will use plastic explosives rather than TNT or dynamite. C-4 and Semtex are two of the better-known types of plastic explosives. They are both formed by mixing an explosive powder with a plastic binding agent to make it less volatile. The plastic also allows the explosive to be shaped any way you want. It can be thought of as Play-Doh for pyromaniacs. Ever seen a Plasticine toy burger explode? Demolition experts pack the explosive into cracks in walls that need to come down. They also use it to cut through steel beams when bringing down buildings.

The blocks of plastic explosives will only explode after being given a large shock, usually from a blasting cap. Without this shock you can actually light them on fire without an explosion. Supposedly you can even shoot a block of plastic explosives and not have it blow up. I think I am just going to trust that fact.

It can take engineers up to three or four months to prepare a building for a three- or four-second collapse. They study the blueprints (if available) to locate load-bearing columns, which are then rigged with explosives and wrapped with fabric to contain the explosion. Other non-load-bearing walls on that particular floor are removed. All of the charges are set to fire in a particular order by using electronic blasting caps – small, electronically operated firecrackers – to begin the explosion of the charge. Interior columns are detonated first to start the building falling inward. A few dozen more explosions and the building becomes a heap of rubble. The outside of the building may also be wrapped in chain link fence or fabric to contain the total explosion. Other buildings in the area are also protected.

Many companies still use the old, easily recognizable T-handle plunger that is pressed down to start the implosion; it just makes for better publicity pictures. The plunger sends an electric charge to the blasting caps and the building starts to fall.

Some companies use the remotes seen in newer movies. These new remotes look like our favourite companion – the television remote. In the high-tech world we live in, some implosions are even started by hitting the Enter key. A computer then controls the firing sequence.

No matter what else is going on, safety is always the number one concern for the blasters. They announce the demolition to all the local news stations for safety and to ensure a big crowd. This gigantic crash is sure to be filmed for the six o'clock news (or maybe for a movie) and will attract a lot of spectators. However, even though getting paid to play with explosives could be a lot of fun, remember the danger and don't forget to wear your hard hat!

Did you know?

Soldiers in Vietnam would actually light strips of C-4 to heat their meals.

How do hand grenades work?

One of the earlier forms of explosives in battle was the hand grenade. Grenades are just gigantic firecrackers designed to kill, maim or disorientate. The term grenade actually comes from the word pomegranate, since the shrapnel reminded people of those wonderful, sweet seeds. Over 1,000 years ago, warring folk filled canisters with explosives and any metal and bits of crap they had lying around to create grenades. Grenades have come a long way since and now come in many different varieties.

Mention 'grenade' and most people imagine what are known as fragmentation grenades. In a fragmentation grenade, a hard plastic or metal shell surrounds an explosive and a percussion cap. A pin is inserted to keep the grenade safe. When the pin is pulled, you are still safe as long as you squeeze the safety handle. When the grenade is thrown, the handle flies off and the timer starts. Most grenades will explode two to six seconds after the handle is released. Upon explosion, the casing fragments into tiny pieces creating shrapnel that can injure and maim.

Grenades have been developed for all sorts of specialized uses over the last hundred years. Smoke and tear gas have both been added to grenades. Smoke grenades create thick, dense smoke used to signal a location or give people or vehicles something to hide behind. Tear gas grenades are designed to fill an enclosed space with tear gas or to create a barrier in front of a rioting crowd.

Several grenades, primarily designed not to kill, have been developed for the police. These include the stun and the sting grenades. Stun grenades, called 'flashbangs' by police, create a tremendous amount of light and a deafening noise. Both of these will combine to completely disorientate an unsuspecting person for up to five seconds. Sting grenades have two rubber shells with tiny superballs placed between the two shells. When they explode, the area is covered with stinging pellets. Nasty!

Some grenades are designed to explode on impact instead of using a timed fuse, which rather limits the ability of an enemy to throw them back. These grenades are safe until activated. Today many grenades are launched instead of thrown. Now you don't have to be a professional cricketer to heave them a long way. Pull the trigger and away they go. After they leave the barrel of the launcher, stabilizing fins come out and the fuse is activated. They will explode on contact. Never knowingly underequipped with things that go bang, the US Army even developed a machine gun that fires 400 grenade rounds a minute.

Grenades are currently being developed that will explode a certain distance away from the gun. This will give troops the ability to injure a person standing behind a barricade.

How do you make butane lighters explode?

Bored? Curious? Or just feeling anti-social? Then try this out on used-up disposable butane lighters. Use the lighter for any and all variety of pyrotechnical pursuits until it's empty. Then stand twenty-five or thirty feet away from a concrete wall and heave the lighter as hard as you can. Upon hitting the wall, some lighters will create a loud pop along with a mini-burst of flame. Some probably won't and you could look really stupid in front of your friends. For safety reasons, you and your friends should keep a good distance away from the wall.

Why does this work? Even after it is supposedly empty, the lighter still contains trace amounts of butane. The lighter will break open when it hits the wall and you also may get a spark. Put the little amount of butane together with that spark and you get a pop. NEVER try this with a full lighter, obviously. It's not supposed to work when a lighter is full because the vapour ignites better than the liquid, but it is not worth the chance.

Did you know?

Butane lighters won't light when they are extremely cold. The butane won't vaporize easily at low temperatures.

How do you start a fire in the woods?

Friction creates heat. Think about rubbing your hands together on a cold day. Rubbing two sticks together can work the same way to create heat and if you're some sort of Ray Mears-style legend (and can rub incredibly fast) you may get a fire. You can also make a bow using string on a stick. Wrap the string around another stick and place the end of this stick on a piece of wood surrounded by dry brush. Then just play the bow like a violinist. With any luck, the friction will eventually light the dry brush.

Obviously, a far better way to start a fire in the woods is to use matches. My motto is 'If you go camping, take matches.' As an adult, camping trips often mean romance and you don't want to wear yourself out rubbing sticks. The stick trick is cool if you can pull it off without looking like an idiot, but matches will get you on your way to a romantic fire quicker. An entire bottle of lighter fluid will help you even more in your quest for fire. But be careful; you'll lose any chance for romance if you burn the tent down.

How does Hollywood create car crashes?

Push an abandoned car off a cliff and you get a twisted lump of junked steel. Turn on a movie camera and repeat and the car will mysteriously burst into flames. Filmmakers have known this secret for years. All of their car crashes involve giant fireballs. My favourite car crashes involve a car going off a cliff and actually igniting on the way down. In real life, a car landing after a fall has maybe a one in a million chance of catching fire. That's because liquid petrol doesn't explode very well; it just burns. Petrol vapour is the stuff that causes explosions and you need a lot of it.

To get the explosions we're used to seeing on screen, stunt co-ordinators cheat by loading a car full of fireworks designed to explode at the director's cue. They have the ultimate pyromaniac dream job; they play with fire and lob cars off cliffs for a living and the bigger the inferno the better.

Are safety matches really safe?

Good old safety matches are designed to ignite only by using a striker board attached to the pack. Striker boards were actually placed inside the cover of the first matchbooks, which caused the entire pack to ignite. Today the striker board is on the outside of the pack. The head of the safety match usually contains potassium chlorate mixed with sulphur, a binder and glass powder. The striker board contains red phosphorus, a binder and powdered glass. Rubbing the match across the striker creates heat because of friction. When you strike a match, a tiny amount of the red phosphorus becomes white phosphorus, which ignites easily with the heat created. This sets offs a decomposition of the potassium chlorate to release oxygen. The sulphur ignites and burns the wooden stick or cardboard strip.

Strike-anywhere matches are hard to find because they are more dangerous. These matches have a different coloured head that contains phosphorus sulphide, potassium chlorate, binders and glass. The phosphorus sulphide ignites easily and the potassium chlorate releases oxygen, which causes the match to burn brighter. The stick will also burn to give you a longer flame. Strike-anywhere matches are misnamed. They don't actually light anywhere, so they should probably be called strike-almost-anywhere-that-is-rough-and-dry matches. You can't light one in a swimming pool or on a dinner plate, but zips, old jeans and the bottom of your shoe all work. I've never tried my fly; there are parts of my body that are scared of fire.

How do they put the fizz into Space Dust?

This pop-in-your-mouth kids' favourite is unique and fun to eat. The candy is created using high-pressure carbon dioxide similar to what you find in soda. The ingredients for Space Dust are melted, the carbon dioxide is added and the concoction is allowed to cool. During cooling, tiny carbon dioxide bubbles are trapped in the candy. As the candy comes into contact with the hot saliva in your mouth, it melts, the bubbles pop open and you get that weird feeling on your tongue.

In the USA, a similar candy called Pop Rocks also gave rise to one of the most enduring urban myths ever – that drinking a cola and eating Pop Rocks would cause your stomach to explode. For this reason, sales dropped and General Foods actually stopped selling the candy, even selling the rights and formula. Luckily, a new company brought the candy back for a whole new generation of American kids.

A neat trick is to place some Space Dust on your tongue and try to hold it against the roof of your mouth. Try not to do this while your boss is watching. He or she might laugh, but you won't.

Scientifically speaking

If you can see light at the end of the tunnel, you are looking the wrong way.

- Barry Commoner

How do you make Polo mints spark?

When chewed in a dark room, Polo mints create an interesting phenomenon: they make your mouth glow blue! Try it for yourself. Go into a bathroom, turn off the lights, let your eyes adjust and crunch away. You get tiny bolts of lightning.

The light is due to a process called triboluminescence. It comes from the electrons that are ripped free as you chew the candy. These electrons combine with nitrogen molecules to produce light. Normally this is light you couldn't see, but the oil ingredient is fluorescent, which helps the process along. The fluorescent oil absorbs short-wavelength light (that we can't see) and emits longer-wavelength visible light. Your mouth glows!

Remember, the light created through this process is very faint, so don't throw out your torches and buy thousands of packets of Polos just yet.

Why do your feet not burn while firewalking?

Walking across fire is frequently associated with islands in the South Pacific, girls in grass skirts, umbrella drinks and coconuts. I just associate firewalking with two words – MENTAL and ILLNESS. I originally thought you had to be somewhat odd – or boozed up – to think this a good idea, but as I studied it, I realized these guys are experts in science. Firewalking is actually a rite of passage, a healing ritual and a test of faith for many cultures.

Firewalking is most often done over a bed of glowing embers and the science is fascinating. When hot and cold objects come together, the hot one loses heat to the cold object due primarily to molecules colliding at the surface. If they stay together long enough, they'll reach the same temperature. How fast they reach this equilibrium temperature depends upon specific heat capacity, mass and thermal conductivity for each object. The product of mass and specific heat capacity for any object is called heat capacity. Energy is conserved, so the heat capacity lost by the coals is equal to the heat capacity gained by the feet.

The final temperature will end up closer to the initial temperature of the feet for several reasons. One, our bodies (and therefore our feet) contain water. Water has a high specific heat capacity and the coals have a low specific heat capacity. This causes the foot temperature to rise only a small amount. Two, the thermal conductivity of water is high. This spreads the heat throughout your body quickly so your soles don't burn (as long as you keep moving). The conductivity of the coals is very low, so heat from other parts of the coals can't get to the top very easily. The coals are also covered in ash, which has an even lower rate of thermal conductivity. Third, the coals are uneven, so not very much of your foot has to be in contact with the coal at any time.

That said, never try firewalking at home (especially not on the living room carpet). There are a few musts that have to be taken into account: The coals must burn a long time to completely remove any water and never under any circumstances should you stop in the middle. Fill up on umbrella drinks and get across the coals as fast as possible. Afterwards, then you can think about chasing the girls with the grass skirts.

Brain fart

Firewalking should never be done while wearing a grass skirt or a kilt.

2. Rev your engines

For men, the need for speed is as basic as life itself. The first caveman that walked the earth learned that speed was the secret to staying alive. He had to be able to outrun the sabre-toothed tiger to see the light of the next day. Well actually, he just had to outrun one fellow caveperson to survive. The first race was born and we've been racing ever since.

The occasional motorway patrol is just part of the game. Blokes simply like to compete. Educational experts have said that competition hurts self-esteem, but these experts are just frustrated because they never won anything. Even a visit to the relatives can be turned into a race by challenging family members to set record times up the M6 (take the toll road, obviously).

Racing technology has grown from racing on the beach to today's super speedways and whether it's Formula 1, drag racing, motorcycle racing, the Tour de France, monster trucks or even lawn mowers, guys love a good race. And racing involves science. So strap in and hold on as we examine the world of racing.

Why don't some racing car tyres have treads?

Dragsters, racing bikes and NASCAR racers' standard race tyres are all made without treads. Tread helps a car maintain traction (grip) with the road surface in wet weather and when turning. Passenger car tyres come with treads for that reason and that's why most races are delayed by the slightest hint of rain.

Tyres are the most important part of a racing car because they are the only part in contact with the road. Normal car tyres are hard, which helps them last longer, but racing car tyres are soft. The soft rubber uses adhesion to increase traction. Walk on a racetrack and you will see thousands of rubber marbles at the bottom of the banked track. These rubber marbles come from tyre erosion and form little balls that roll to the bottom. Racing car tyres are also wider, which gives the driver even more grip since more rubber adheres to the road. The grip of the tyres is what helps the cars turn the corners at such speed and g-loading.

Racing car tyres are designed to be driven only a few hundred miles and are specially designed for each track. In the American motorsport of NASCAR, the tyres are even designed specifically for each side of the car. On most NASCAR tracks where drivers make only left-hand turns, the inside and outside tyres undergo different stresses. Formula 1 racing currently requires each tyre to have four treads around it, but those rules might change in the future. They are required to have treads to decrease traction, which slows the cars down. Formula 1 cars are the only major racing car type to race in the rain. Of course, they change to more heavily treaded tyres before resuming the race after it begins to rain.

What is the advantage of nitrogen in tyres?

Most racing car crew chiefs fill their tyres with nitrogen, as do many truck drivers, bike racers and the United States military. Using nitrogen in tyres is even creeping into the everyday-driver market. Nitrogen is relatively inert, especially when compared to oxygen. It's readily available and, therefore, relatively cheap; it makes up about 80 per cent of the air around us. Nitrogen is dry out of a bottle, so no moisture gets inside the tyre. Moisture in the tyre vaporizes at high temperatures and expands, which would adversely affect the handling of the car. Nitrogen also migrates through the wall of the tyre at a much slower rate than regular air and best of all, it runs at a cooler temperature in a tyre than regular air. All of these make nitrogen the choice of most racers.

If nitrogen is so great, how about filling your tyres with helium? Your car/bike could just float down the road hardly touching the ground. If helium works, hydrogen would be even better since it is the lightest gas. Of course, helium and hydrogen are flammable, so at a certain speed your tyres would explode. Better stick to nitrogen.

How does downforce help racers on the track?

To understand downforce, we must first understand upforce (lift) on aeroplanes. Hold a ten pound note or a strip of paper under your bottom lip and blow straight across the top of the paper. Why it rises is explained by the Bernoulli principle, which states that as the velocity of a fluid increases, the pressure decreases. Above the note you have a lower pressure than under the note and lift is created. The aerofoil of a wing creates lift for an aeroplane because of its shape. The air passes faster over the top of the wing so there's a lower pressure above the wing. The difference in pressure above and below the wing causes the wing to lift.

Racing cars have inverted wings attached to their frame. By tilting these wings, the pit crews can help control the amount of downforce, which helps to increase friction with the road. Friction is desirable for traction and cornering. More friction means your tyres won't spin and that you can drive faster through the corners.

Most single-car crashes occur at corners because drivers are going too fast. Driving too fast in a corner means the centripetal force is greater than the friction, so the car slides. The car will slide tangential to the circle and into the wall. At most racetracks you can see wall marks at each corner from crashes. However, most multicar accidents occur when one racer hits another racer. Men like racing, but they absolutely love a good crash, especially since the cars are so safe nowadays.

Racing car manufacturers do all that they can to keep their drivers safe and that includes using the entire body of the car to help create downforce. Formula 1 cars have bodies that are actually shaped like upside-down wings. The narrow area under the car creates high-velocity air under the car, which lowers the pressure so the higher-pressure air above the car pushes it down at high speeds. NASCAR uses air dams to limit the amount of air passing under the body of the vehicle for the same reason. Aeroplanes actually take off at lower speeds than most high-end racing cars!

Dragsters have an interesting way of creating more downforce: they point the exhaust headers to the sky. The exhaust coming out pushes the car down into the track. More than half of their total downforce comes from this. It also creates a crowd-pleasing sight and a deafening noise. The exhaust gases will often light after they leave the headers for extra pyrotechnics. The dragsters also have a wing over the back wheels to generate more downforce.

Did you know?

Drag racing got its start on the famous salt flats in California's Mojave Desert and gained popularity during the Second World War.

Why do dragsters do burnouts?

Dragsters go through a burnout prior to entering the staging area. This cloud-inducing ritual is always a hit with the grease-monkeys in the crowd. We get noise and a ton of smoke. However, in addition to amping up the crowd, the burnout serves a scientific purpose.

The dragster pulls into a puddle of water to wet the tyres down. After pulling onto dry pavement, the brakes are locked and the throttle is opened up. The back wheels spin in place and create smoke. The tyres also grow and expand as they heat up. Like most racing car tyres, dragster car wheels are soft rubber. The expanded diameter helps the dragster cover more distance with each rotation of the tyre. The burnout makes the tyres stickier and cleans off any debris. It also wears out the tyre quickly, but the top-level pros get them for free. The cars are equipped with a line lock, which allows the back brakes to be disengaged during the burnout, saving the back brake pads.

The tyres are also only filled with 7 pounds per square inch (psi) of pressure. A normal car tyre is usually filled to about 30 psi and bike tyres can be up to 100 psi. The low pressure causes the tyre to look flat. The tyres are actually called wrinkle-walls because of this look. The flattened tyre allows more of the tyre to form a larger contact patch with the ground. A greater contact patch leads to greater traction.

What is the difference between nitro and top fuel?

Nitro is short for nitromethane, a popular fuel added into engines for increased horsepower. All hydrocarbons burn, but nitro (CH_3NO_2) has the added benefit of containing oxygen. That means it can burn using less outside oxygen. It delivers almost two and a half times the power of conventional petrol. Nitro also vaporizes at a lower temperature, so the engine block is actually cooled by the nitro. Outside oxygen is usually the limiting factor in an engine's horsepower because it controls how much fuel is burned. Nitro helps solve that problem.

Nitro is used in street racers. It is also the largest component of top fuel that powers the long-rail dragsters. Top fuel is 85 per cent nitromethane and 15 per cent methanol. Top fuel burns fast and ferocious, just what redneck speed freaks want.

Brain fart

If nitro is so good for racing, wouldn't it be great for my gas-fired barbecue? Just wait until everything else is on the table, kick in the nitro, and the steaks and sausages will be done. Of course, you also run the risk of launching your barbecue off the patio. Slow-done steaks or a barbecue in orbit? Up to you.

How do 'smart walls' work?

As safety continues to rise to new levels, smart wall technology has become standard on most racetracks. Racing car fatalities are now rare on the professional circuits due to this new technology and better head restraints. The head restraints redirect much of the initial energy away from the head and neck by keeping the movement of the head and neck to a minimum. New wall technology only adds to the safety factor.

The outside concrete wall is now backed up by at least two inner layers. The innermost layer is composed of hollow steel beams that are strong but will crush in an extreme crash. The inner layer is made up of very high-density foam. In a head-on crash, the two inner layers act like a giant sponge to absorb the energy. The more time it takes to absorb the energy, the less force the driver feels. Safe-wall technology, although new to racing, has been used on many motorways for over twenty years.

Most racing car crashes are glancing blows, which decreases the amount of force on the car and driver. Also, cars are designed to crumple only at certain points. Manufacturers just need to protect the cockpit area, called the driver's capsule. Many people equate racing with good old-fashioned brass neck and balls of steel. Well, maybe so, but the gutsy drivers are backed by boffins with engineering degrees. Behind every good racing driver is a calculator-carrying geek. Never pick on such folk. They may save your neck one day! And hopefully your balls while they're at it...

How does a Stolen Vehicle Recovery System work?

If you drive a nice car, people will be jealous. Some may even try to steal it. A stolen vehicle recovery system works by hiding a GPS receiver (or computer chip) in or on your car. After the vehicle is stolen, the system can be activated to find the car's location. You will have to pay for the system's installation and a fee to monitor it. The receiver must be well hidden or the thieves will just disable it or nick it and sell it for a profit too. A cheaper stolen vehicle recovery system is buying a car no one would ever steal. A well-rusted 1991 Toyota Corolla may even be cheaper than the cost of a system. The added benefit of this method is you will get extra exercise. You're guaranteed to park your pride and joy out of the sight of your friends and colleagues.

How does GPS work in your car?

Blokes are important to the future of the car industry. After all, if not for the male refusal to ask for directions, GPS-aided navigational devices wouldn't be standard in all decent cars these days. These marvels of technology help us on many levels. The fact that we don't have to stop the car to ask directions goes without saying, but we also get a computerized female voice telling us to turn right in three hundred yards. Most blokes can't stand backseat drivers, or even passenger-seat navigators, but we have no problems taking directions from that voice. Some fellas even find it sexy. Depending on the type of system, we also get a built-in television screen in the front seat. Guys would never watch television while driving, but it's nice to know we can watch it while waiting to pick up the kids from school.

And if you don't want to be ordered around by that sexy female voice, you have other options. Most units come with a choice of different voices. At least one manufacturer allows you to go online and choose 'celebrity' voices. I can almost hear Katie Price's nasal voice intoning 'Turn right or whatever'. Or Yoda's more contemplative 'A tremor in the force I feel' as you take the wrong turning.

GPS stands for Global Positioning System and was developed for the military, but, luckily for blokes, it was soon opened up to the general public. This system uses twenty-seven satellites orbiting the Earth which transmit data to receivers on the ground. They say that every square centimetre of the Earth's surface is visible to at least four satellites at any one time and your GPS receiver calculates the distance to at least three (some units use four) of these satellites. From these different distances, the receiver uses a process called trilateration to combine these distances with the location of the satellites to determine the location of the receiver (your individual GPS unit). Think of the distance from one satellite as the radius of a circle. Another distance would be another circle. Those circles would only overlap at two points. A distance from a third satellite would give you the exact location of the receiver. Of course, in three-dimensional space the circles are actually spheres, but the concept is the same.

The location of your car can then be fed into the unit's database of maps and the sexy/fake-tanned/green voice will point you to your destination. Many devices also pre-program locations of petrol stations, restaurants and cash machines into the database. A few databases even contain locations of known speed cameras to help save you a few points on your licence. These receivers are also useful for automotive security if your car gets stolen.

Did you know?

Former US President Ronald Reagan was responsible for opening up the commercial GPS market. After the shoot-down of the Korean Air 747 in 1983 (for inadvertently entering Soviet airspace), Reagan issued a directive that the GPS signals would be freely available to the world when the system was ready.

How do antilock brakes (ABS) work?

If you are old (like me), you were taught to pump your brakes as you slowed. This was to prevent the wheels from locking up and your car from sliding uncontrollably. ABS electronically pumps your brakes, so none of the wheels should ever slide. The best ABS systems use a sensor on each wheel to measure the rotational speed for that wheel. A sliding wheel does not rotate. As the sensor detects one wheel rapidly slowing (compared with the other wheels), a valve will shut off brake pressure to that wheel. This causes the wheel to keep rolling. The valve reopens when the wheel is spinning at the same rate as the others. The valve may open and close up to twenty times per second. The wheels keep spinning and the car is easier to control as it comes to a stop.

In early ABS systems, you could actually feel the brake pedal pumping under your foot, but with newer systems this is no longer felt. You should never pump your brakes with ABS; obviously this defeats the object and it will just take you longer to stop.

The science of ABS was learned by most of us on a bicycle many years ago. If you somehow managed to lock up the wheels, you crashed into the side of your Dad's just-cleaned and polished car. By steering as you 'stopped', you could instead plough into a bush in the neighbours' garden. Crash landing into the bush was the bicycle equivalent of the airbag.

Can you drive on flat tyres?

Run-flat tyres – tyres that will keep rolling even as they lose air (or nitrogen) – are becoming more common. Newer technology can even handle blowouts with minimal loss of control. Run-flat tyres use one or a combination of three separate methods. Self-sealing tyres use a fluid that will seal any small punctures. The fluid reacts to the loss of air and plugs smaller holes. This fluid is actually sold separately in aerosol cans or is inside the tyre originally. Self-supporting tyres have very rigid sidewalls that support the entire car for short distances. Auxiliary supported systems contain a ring of rubber and/or polyurethane inside the traditional tyre. Think of them as having a second tyre waiting as backup within the outer tyre. When air is lost, the car rides on the inner ring. These auxiliary supported systems are the surest handling form of run-flat tyres.

Actually the best run-flat tyres are on my lawn mower. I've never had a puncture in twenty years of mowing lawns. You could just put big solid wheels on your Fiesta and trundle down the road. Of course, your fillings would shake out of your teeth, but you'd never be stranded with a flat.

When does talcum powder punch you in the face?

For years, when I thought of the word airbag, my secondary school art teacher came to mind. Today airbags take on a new meaning. We are discussing the powder-exploding-in-your-face-save-your-life car airbags. Airbags save lives because of physics and chemistry. The physics behind an airbag is the same as trapping a football. All moving bodies contain momentum (the product of mass and velocity). To stop an object, you need to apply an impulse (the product of force and time). It takes the same impulse to stop whether you hit a tree or use the brakes. The key is time. If you increase the time, you decrease the force on the object. The airbag increases the time it takes your body to stop, so you feel less force. In the same way, moving your foot as you trap the ball decreases the force the football (and your foot) feels, keeping the ball under closer control as it stops.

Airbags contain an accelerometer that measures rapid deceleration, like when you hit a tree. This accelerometer activates the airbag's chemical reaction to inflate the bag. The most common chemical used is sodium azide. When ignited, it releases harmless nitrogen gas into the nylon bag, very fast. The bag deploys at speeds up to 220 mph! Your body must be 10 inches away from the bag when it explodes to avoid serious damage. The bag increases the time your body takes to stop and you feel less force. The bag also spreads out the force over a larger portion of your body, which helps spread the hurt out. The bag immediately starts to deflate the nitrogen through vents so you can get out of the car. Of course, you are left covered with talcum powder and maybe some fabric burns. This powder is used to lubricate the bag while the bag is folded up in your car. Think of it as being punched with a giant, padded, baby-powder-covered boxing glove. Better that than the alternative, eh?

Today many newer cars come with side airbags. Car scientists are experimenting with selective airbags, which deploy at different speeds for different crashes. These airbags will also deploy at different speeds based on where you are sitting on the seat. Remember, airbags should always be used with seat belts and children should never be in the front seat. They are too small and could be hurt by the airbag.

You may be wondering, 'If airbags can help me survive, why can't they help my car survive?' Most men would love a car that had airbags mounted on the outside. When we start to hit something our car would explode into a balloon-covered ball of steel and nylon. We could bounce around like a carnival ride and our car would survive unhurt. James Bond once used a business suit like this to survive. NASA also used a similar setup to land a spacecraft on Mars. Why not use this same technology to save our cars?

Did you know?

Honda introduced the first airbag system for motorcycles in 2006.

Can you make your own petrol?

With the fluctuating price of petrol, you might wonder if it's possible to make your own. Well, you can use a process called gasification that utilizes organic waste to make syngas (or synthetic gas, which you can then use to make petrol), but it is easier to make ethanol that can power your car. Ethanol is just a simple form of alcohol that already makes up 5 to 15 per cent of the petrol we buy. So how do we make ethanol?

To make alcohol, you need a sugary/starchy food and some yeast. For the sugary/starchy stuff you can use corn, but corn is average in its starch content. Sugar cane is a much better choice. Yeast is living bacteria that devour the sugar or starch in your food. The yeast eats through the starch/sugar and excretes alcohol, water and carbon dioxide in the fermentation process, leaving only alcohol and water. This is where your local bearded home-brewer will come in handy. He probably has a still in his loft or garage. The still separates the alcohol from the water, so you are left with pure alcohol. Alcohol evaporates much easier than water. By heating the mixture, the alcohol is vaporized. The alcohol condenses and runs through a tube to another collecting device (we'll call it a jug). The water is left in the main kettle. The collecting jug will be full of pure alcohol. Pour it into your petrol tank and away you go. Be sure to thank The Beard as you drive away. Note: to use pure ethanol in a car engine, you'll need to get hold of an ethanol conversion kit.

Did you know?

Most people think of 'moonshine' as illegal alcohol from the southern states of the USA, but the word actually originated in Britain. Smugglers worked by moonlight as they brought in illegal and contraband goods, including alcohol. What they were smuggling became known as moonshine. The word spread to America in the mid-19th century and was used to describe 'liquor' made to avoid taxes or prohibitionists.

How do you hot-wire a car?

The concept of hot-wiring a car is fairly simple on older models. The starter, battery and the ignition switch are all in a simple electrical circuit. The key allows you to close the switch to complete the circuit. You use the same process when you flick a light switch. To hot-wire a car, you simply cut the two wires going into the ignition. Strip back both wires a little, twist them together and the circuit is complete. The engine roars to life. Of course, movies never show the guy getting shocked as he twists the wires together. Trust me; it would shock the living daylights out of you. It's never a good idea to twist together two wires in a live circuit, unless you are demented enough to like pain.

What is the secret behind digital chip keys?

Newer cars are harder to hot-wire because they come with keys that contain a radio frequency transmitter. The key sends out a radio signal to your car that allows your key – and only your key – to start your car. Think of it like a second switch that's necessary to complete the circuit. The radio signal closes one switch and the key closes the other. Some models even scramble the radio code every time the key is used. This doesn't mean your car can't be hot-wired. Actually it could be hot-wired twice, once around both switches, but I don't think they teach the double hot-wire in acting school, yet.

A few posh cars don't even need an ignition slot at all. The key stays in your pocket and you just press a button on the dash. The key transmits a radio signal that closes the circuit. Racing car drivers just press a button, which is a shame because it would be funny to see Lewis Hamilton throwing a fit at the Monaco Grand Prix because of a lost ignition key. You can even buy fingerprint readers that allow you to start your car without a key. Then the thief has to steal your finger to steal your car. Hopefully that counts as a step too far.

What's the difference between torque and horsepower?

No argument arouses more grease-stained passion among petrolheads than the exciting old chestnut 'torque versus horsepower' issue. The fact is, they are related. Torque is a twisting force that doesn't require motion. It is measured out of the engine at the crankshaft by a machine called a dynamometer. Horsepower requires movement and is a measurement of the work and power an engine delivers. The unit 'horsepower' was actually just a clever marketing gimmick to help canny old Scottish engineer James Watt sell his new steam engines in the late 1700s.

Car magazines list both torque and horsepower at certain revolutions per minute (rpm). At low rpms, torque is more important to get you moving. At higher rpms, horsepower is more important to keep the wheels spinning. Your transmission uses gears to maximize the power delivered to the road. The combination of torque, horsepower and gearing is designed for specific needs. A farm tractor has incredible torque but low rpms. The tractor just crawls along, but it pulls easily. A high-performance car has incredible horsepower at high rpms but may have less torque than the tractor. For this reason, you never see a sports car pulling a plough.

In their love for speed, men will race tractors, but they race tractors against other tractors. The sight of a Massey Ferguson or a JCB coming off the final corner in the next British Grand Prix would be one to treasure. Blokes will continue to debate torque and horsepower, but the truth is they are both important for different needs. When taking off from a red light, torque rules. But on the Autobahn, horsepower rules. Especially if it belongs to a German police BMW...

Scientifically speaking

All truths are easy to understand once they are discovered; the point is to discover them.

- Galileo

3. Game time

For many of us, sport represents a way of life. Even if we are only average players, your average bloke revels in watching sports done well, whether it's witnessing Premiership stars dazzle us with their fancy footwork or watching Tiger Woods do things we never will (with a golf club, I mean). We secretly wish we could get paid millions to play a kids' game. We watch, we cheer, we go out and try to imitate them, yet most of us are basically too unfit or lazy or talentless to get much further. But science is helping us be less crap than we were before.

The science of sports has grown a hundredfold over the last fifty years. Computers are no longer the tools of just nerds. Engineers have turned their sights to round balls, bats, shoes and all the things we use to make sport fun. Even things as mundane as clothes and bowling balls have been engineered for better results.

Sports-related items are a multibillion-pound business. The high-tech science types spend tons of cash to develop ways to make our game better. It is funny that we buy into the mystique of fancier equipment. Most of us won't be professional players with fancier stuff, but we buy anyway. We gladly pay extra money for any piece of equipment that will help us beat our mates in a game. So lace up your trainers as we examine the wonderful world of sports.

Can a graphite shaft help you hit a golf ball further?

Most of us who play golf waver between a love of the sport and a hatred of it. We spend a soggy Saturday alternately enjoying the good walk and swearing like a trooper, but most of us keep coming back to the course. Almost all golfers fall into three categories: pros, scratch golfers and hacks. Pros make a living playing golf. Scratch golfers are good but don't make a living at the game. Hackers hack. The ball seldom goes in the planned direction when a hacker connects.

I think hackers get more enjoyment out of the game than the professionals. Before you put down this book, let me defend my point. You pay a certain amount for greens fees and a few other things. Amateurs don't get a discount for using fewer strokes, so you get more bang for your buck. Hackers also get to enjoy all the parts of the course, including the back gardens of a few people who live close by. Of course, a few amateurs want to be good so they invest a lot of dough into getting better.

We copy the pros and buy the best we can afford, but many times the clubs we buy – our largest financial investment in the game – aren't the best for us. A golf swing is all about transferring energy into the ball. Pros swing very fast and true and generate tons of energy from the speed of their swing. Most pros use a very stiff shaft so little energy is lost to flexing the shaft and vibrations after the ball is hit. Amateurs need a little more help.

When hitting the ball, speed is king because the primary energy is kinetic and kinetic energy is related to the square of the speed. A faster clubhead delivers more energy to the ball and the whip action a flexible shaft generates helps to create additional speed. As you get better and win a few quid from your friends you can upgrade to stiffer shafts, but you might not get to see as much of the course.

Graphite, carbon fibre and steel shafts can all be tailored to your game. Graphite is a good choice for many of us for two reasons: one, it is light so the club can be swung faster; and two, graphite is more flexible than steel so you can get good whip action. However, even though graphite can help you hit a ball further, the increased flexibility of the shaft is harder to control so you may sacrifice accuracy for distance. A personal note – all shafts will break when you wrap them round a tree in anger.

How do large golf clubheads help?

Large heads (for the golf club, you moron) result in longer drives. The key here is energy and momentum conservation. Titanium and exotic alloys help to make clubheads larger and lighter. A lighter clubhead can reach a higher speed and a higher speed means more kinetic energy for the club. Additional energy for the club equals more energy for the ball. The ball goes further.

A larger clubhead also gives you a larger sweet spot – the ideal place to strike the ball – on the clubface. The actual contact with the ball is all about momentum transfer. Momentum is the product of mass and velocity and is of utmost importance in collisions. A more massive clubhead gives you more momentum to deliver to the ball provided your swing velocity is high enough. Heavier clubs will slow your swing down, so you walk a fine line between mass and velocity.

At least in golf, bigger is better. But too big is not worth the lost clubhead speed. You actually want to crush drives off the tee. Your buddies all watch as you tee it up. You can walk a little taller and stick your chest out further when you are long off the tee. And remember this rule of golf: If you don't reach the red tees, you must stick something else out.

Can three-piece golf balls help your game?

Balls are important in life and also in golf. Once you reach a certain level with your golf game, the ball can help take you to the next step. Ball scientists have spent years developing balls that will help your game and generate their company piles of cash. Golf actually started with players swinging at feather-filled sacks. We've come a long way since then.

The newest generation of balls contains three layers for increased performance. The thin, dimpled outer cover is soft, which allows for increased spin. The additional spin is especially important around the greens. The next layer is a hard mantle that helps transmit the club's energy into the core. The core is hard and designed for distance. Put it all together and you have a great ball for pros and scratch golfers.

For hackers, though, three-piece balls are probably not going to help. The increased spin will contribute to slicing and hooking and will cost you distance if your swing is slow. The soft cover will also tear if you mis-hit the ball. The ball will be left with a giant smile in the torn cover. These balls are expensive and will sink in the water. Most hackers could save a few quid by playing floaters stolen from your favourite aquatic driving range. You also get an added benefit if you actually ever beat anybody with these balls. Can you imagine the embarrassment of being thrashed by a player using a ball with a red stripe? You are sure to be treated well at the nineteenth hole if you promise to keep the secret. The thrill of victory and free drinks.

Why do they put dimples on a golf ball?

Early golf balls were smooth. Golfers soon realized that a scuffed-up ball would actually fly further. Many a golfer would take a little sandpaper to the ball to lower his score. Dimples were added and the rest is history.

There are two types of aerodynamic drag on any flying ball: friction and pressure drag due to air separation. Friction with the air is a given unless you play on the moon. Pressure drag due to air separation can be lessened if you can get the air to curve further around the ball. Dimples do just this by grabbing boundary air and causing it to stay with the ball longer. The longer the boundary air stays curved around the ball, the less pressure drag there is. Less drag equals longer drives. Personal note from my twenty-year love-hate relationship with the game – dimpled golf balls still sink.

Why are tennis balls fuzzy?

Fuzz on a tennis ball serves the same purpose as dimples on a golf ball. The fuzz creates small swirls of air very near the surface of the ball. These small swirls stick to the ball as it spins. The small swirls are called the boundary layer. Air will slide over these swirls of boundary air easily.

The balls will still curve to the Magnus and Bernoulli effects even with this boundary layer of air. Tennis balls are usually hit with plenty of topspin. A tennis player strikes the ball in an upward arc as the racquet goes from low to high. The racquet imparts topspin.

This topspin causes the ball to curve downwards, so you can hit the ball hard and make it curve back down onto the court. Bernoulli and the boundary layer help you to be the best. And we owe it all to a little fuzz (and some science). Imagine if Magnus and Bernoulli ever played against each other... New balls please!

Why are new tennis racquets so massive?

Gone are the skinny wooden racquets that dominated the game for years. Of course, they dominated the game because they were the only option. Starting in the 1970s, racquets began to grow to enormous proportions. The average tennis player wanted to get better and science stepped in. Advances in materials and the growing popularity of the game were primarily responsible for this growth.

Larger racquets have larger sweet spots. But even the sweet spot has several parts. Close to the top of the sweet spot is a dead spot. This spot will transmit maximum vibrations to your arm but minimize the force on the ball. This is the perfect area for hitting a drop shot. Closer to the handle is the maximum bounce spot, giving the ball a maximum bounce for a minimum swing – but when you're swinging the racquet it is better to use more of the racquet. Between these two spots is the centre of percussion. This area gives the least vibration to your arm and is the best overall spot to hit the ball since there is very little wasted energy.

The racquet is basically a long lever and mechanical advantage is gained as the lever becomes longer. Think of it as opening a hinged door from the wrong side. It is very difficult to push open at the hinges. The farther away from the hinge you push, the easier it is to open the door. You can get different handle lengths for additional power, but a nice midsize racquet is good for most of us. The pros may benefit from smaller or longer racquets, but most of us won't notice the difference.

Why does only the cue ball come back in a game of pool?

When it comes to pool, one single question has perplexed blokes of all ages and intellects: how do they get the cue ball back in a coin-operated pool table? All of the other balls go into the hopper and won't be released until more coins go in the chutes, but the cue ball magically reappears after every foul pot. And some of us foul quite a few times in a game. A purely anecdotal science hypothesis is that the number of accidental pots goes up as the pints go down.

Early coin-op tables used heavier and larger cue balls. This method is still employed today for many tables. The cue ball is about ⅛-inch larger in diameter. As it rolls into the collecting chute, it strikes an overhead rail that deflects it into the cue ball opening. Simple and effective, but many good players don't like the larger ball. Newer tables use a cue ball with a magnet (or a metal core) to divert the cue ball. The ball is identical in size to the other balls but is usually heavier. Pool enthusiasts also claim the ball affects play. Magnetic balls also break easier. A few new machines use a laser to measure the amount of light reflected off the white cue ball. The white ball is kicked out before it reaches the collection trough. The white ball is exactly the same size and weight as the other balls, but chalk on the cue ball occasionally fools these sensors.

The average coin-op pool player plays when he's drinking and as a result he's not as good as he thinks he is – but it's easier to blame the ball. You can also blame the lighting, what's playing on the juke box, your mates distracting you and anyone else to hand. Just don't start going on about the internal makeup of the cue ball as you'll sound like a sore loser – after all, the bloke who beat you did so using the same balls.

How do sports clothes wick away sweat?

We sweat to keep our body from overheating during times of exertion. The layer of water on our skins cools us as it evaporates, but if it gets trapped in our clothes, we end up with a sweat-soaked shirt. In the good old days, cotton was the fabric of choice for working out. Cotton fibres are very good at wicking water away from your skin, but the sweat then gets absorbed into the cotton. The shirt will weigh an extra five pounds after a few minutes of a particularly hard workout.

New fibre technology and design allow us to sweat without the build-up of water in our clothes. Engineered fabrics are created by virtually all of the main fabric manufacturers and are sold for performance clothing, bed sheets and even nightgowns for menopausal women. Most of these engineered fabrics are constructed using similar design features.

Firstly, the individual fibres don't absorb sweat the way cotton does. Secondly, the fibres are woven in such away to allow wicking channels to draw moisture away from the skin. And thirdly, the exterior of the fabric is coated with a chemical that attracts water. The sweat is pulled up away from the body and the airflow through the fabric allows the sweat to evaporate faster.

With sports clothing being a multibillion-pound business, we are guaranteed to see further advances every year. These clothes help make sports better for us amateurs who begrudgingly pay too much for engineered clothing that we think helps improve our game. Engineered fabrics are even starting to show up in smarter clobber. In the future, an engineered suit may help you sweat your way through your next job interview.

How can a swimsuit help you swim faster?

Engineered fabrics don't just help you shed sweat; they can help in other ways (aside from preventing you having to play sports in the nude). What about a swimsuit that repels water? High-tech swimsuits for professionals do that and more.

By repelling water, the suit won't absorb any, so it stays as light as possible. The fabric is similar to the fabric used in high-tech anoraks (rain gear, not scientists). The holes in the fabric are too small for water molecules to penetrate. They are also coated in a chemical, similar to Teflon, which makes the water slide off. The swimsuits are also made with welded seams, which are thinner than a traditional sewed seam. The flatter seam allows less drag.

The suits also compress the entire body which means less drag. The suits cover swimmers from their shoulders to their ankles and have much less drag than shaved skin. They will eventually cover the arms and head as companies do more research. Currently swimmers still wear a cap and keep their arms bare, but that will change as science helps us swim faster.

Brain fart

We're not talking about *Borat*-style mankinis here; we're talking about the full-length body suits that world-class swimmers wear.

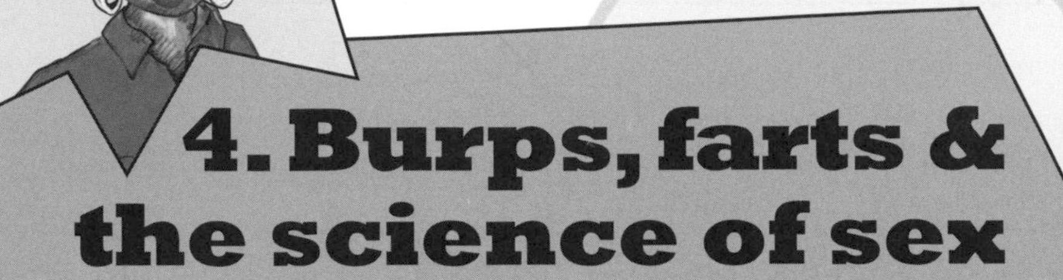

4. Burps, farts & the science of sex

The human body, brain and all its appendages might be the most complex system ever made. The ability to think, reason and react in a fraction of a second is absolutely amazing. It is estimated that a computer would need a thousand times the calculating power of a supercomputer to even do half of what we do every day. So that means at two thousand times the calculating power, they could replace us. With the way computers improve by leaps and bounds, we may be obsolete in a few years.

Blokes spend many hours admiring the human body – our

own and others'. We may not know all of the inner workings, but we enjoy the sights, sounds and smells that are created by this marvellous machine. The human body and the things that make us unique will never be completely replaced by machines. Could you imagine some nerd-boy sitting in a lab trying to get his computer to fart? Actually I can and he would be laughing the entire time.

Women do not understand our fascination with bodily functions and they never will. That is one of the great things that make us different. Of course, men don't understand the need for pedicures, either. So I suppose we are even. Anyway, let's examine all the things that guys need to know about the human body. We need a starting point and it may as well be the colon. The colon is a man's best friend. We all laugh, joke and share the wealth of all things related to this great body part. So put your gas mask on as we examine the wonderful world of the male body.

How are farts created?

Farts cause males to laugh. We may be a crude lot, but we do know what funny is. Even the word itself makes me laugh. However, there are many other ways to say the word fart. Passing wind makes the most sense. Flatulence is the clinical term used by physicians for excess gas in the intestines. So maybe farting should be called flatulating. Cutting the cheese makes no sense to me. I have eaten stinky cheese but nothing that resembles a good guff.

Farts are created by gas in the intestines. This gas comes from a variety of sources. We swallow air as we are eating or drinking. Carbonated drinks help this process along. Gas also seeps through the intestine wall into the food we ate. Gases are also created as we digest food and by bacteria that live in our intestine. Next time you complain about your neighbourhood, just think, you could be living in an intestine. Makes your neighbourhood seem a bit nicer, doesn't it?

Farting ability differs from person to person. A certain relative of mine (who shall remain nameless) is a legend for his farting prowess. He can fart on cue at all times of the day and night. He would be world class at fart poker. For the uninformed, fart poker is a game we played at university. When one person farted, the rest of the room had a minute to ante up and pay a quid to the farter, unless they could see the fart and raise the bet by farting themselves. Then they get the pound by seeing the fart, unless they were raised again. The problem with being a great farter is that you can't turn off the gas pump. They pop out at the most inconvenient times. Of course, if you are adept at controlling the sounds, you can sneak the blame onto unsuspecting friends.

Why do farts make a sound?

The ability to control the sounds emanating from your anus is a skill, a skill that will be admired by all of your male friends. Farts make sounds by a process similar to the reed in the mouthpiece of an oboe. The vibration of the opening creates the sound. Opening is a politically correct term for bum hole.

The sound can take many different forms. Your sphincter muscles control the tightness of your anus. The tightness allows you some control over pitch. A tighter opening equals higher pitched poots. High-velocity gas can also lead to a higher pitch. The sounds range from SBDs (silent but deadly) to rafter-shaking bass notes. Timbre (quality) of the sound is related to the quality of wind, the size of your arse cheeks and the clothes you have on. All three go together to determine tonal quality.

Passing off your breaking wind as someone else's is a truly admired guy skill. I taught with a bloke who was very adept at sneaking a fart off in class. The pupils would invariably blame someone else, never the offending teacher. He really was a truly fabulous teacher – great at leading class discussions with the ability to add a dose of levity when needed.

How do you light a fart?

Of course, it's possible to light a fart. Many farts contain methane and hydrogen, both of which are flammable. The question is why you would want to. Open flame dangerously close to your private parts is just not very clever. Usually fart-lighting episodes occur among people who have had a little Dutch courage, but not always. Some guys deliberately set out to light the gas just to see if it is possible. Never in recorded history has there been a documented case of a woman attempting this cunning stunt.

This is an extremely dangerous stunt and should only be tried by trained professionals. You need the ideal setting and safety clothing. We are talking state-of-the-art flameproof suits like the Stig wears. You could even write letters to all of the people on the *Sunday Times* Rich List requesting sponsorship. Not sure what sort of reaction you'd get, but worth a go.

A doctor's exam table complete with stirrups is the ideal place for fart lighting; even better would be a birthing chair. A cauliflower-cheese-broccoli omelette may help the process along, so stop at your favourite greasy spoon on the way to the doctor's office. Get comfortable on the table with your feet up. Hold a lighter under your leg directly in front of your anus. Yelling 'There she blows!' as the lighter burns will help the process. Let 'er rip and watch the flare-up.

Take note of the colour to indicate different trace gases. Methane burns blue. Sulphur and sodium will give yellow hints to the flame. Copper traces will burn green. Potassium will give purplish tints to the flaming gas.

Seriously, burning farts is dangerous and should only be tried by professionals. A better way to see this is to go to one of the video-sharing websites and type in 'burning farts'. You get to see other loonies risk their anal beards for your enjoyment.

Brain fart

Fart lighting should only be attempted by trained professionals. Wonder where you get that degree then?

What's a 'snart'?

A 'snart' is a combination of sneeze and fart. These usually happen on a particularly windy day when you sneeze. The muscle contractions of your sneeze forces gas out the other end at the exact same instant. Although momentarily painful, there are no reported deaths ever from a snart.

While an occasional fart makes us laugh, snarts never do. Both are natural processes that are better left separate. Funny thing, there is very little research that I can find on snarts. If you know any good PhD candidates looking for a thesis, snarts need to be delved into. We have already discussed farts, so let's take a look at sneezes.

You feel it coming, the tingling. It's coming and you can't stop it. You hold up a hand to your friends and take a step back. You rush to cover your nose with the other hand. Your nose explodes in a violent hailstorm of mucus and sound. Your friends bless you because of the mistaken notion you were momentarily close to death. You shake your head in dismay and if you are like me, prepare for another. I have never single-sneezed in my life.

The nose is a mini air purifier. So when viruses, bacteria, dust, smoke or other allergens enter your nose, the sneeze helps to clear the air. The sensitive linings of your nose send a signal to your brain to get rid of the offending material. The signal goes from the brain and then to the muscles and all of this takes time. We feel the tingle and we know it is coming.

The muscles of the lungs contract, the mouth is sealed off, the vocal cords close and your body spasms. That spasm causes virtually every muscle in your body to contract to get rid of the offensive material. Sneezing is an incredible abdominal workout.

The wonderful sound of a sneeze is brought out by the rapid rush of air as it leaves and your mouth opening to allow more air in. The sounds run the gamut from golf clap to pneumatic drill strength. Most of us have at least one friend whose sneezes are hilarious. I have a mate who goes from burly man to Celine Dion in the flash of a sneeze. I now realize that I am helping him if I can make him sneeze. He gets pure air and his belly gets a much-needed workout.

Did you know?

People have been saying 'Bless you' after a sneeze since AD 77. It was believed that the heart stopped during a sneeze, and blessing the sneezer ensured that the heart would restart and the person would stay alive!

How does erectile dysfunction work (or not)?

Erectile dysfunction. The two words together just make blokes snigger... well, at least some blokes. And I have never heard a woman laugh at those words, only men and probably nervous ones at that. I am not saying women don't laugh, I've just never heard them. But ED isn't funny. Much.

To understand erectile dysfunction (ED), we first have to understand erectile function (EF). An erection is as simple as blowing up a balloon. Normally the arteries going into ~~Mr Happy~~, ~~the old chap~~, ~~King Kong~~ your penis are somewhat constricted and the veins are wide open. Virtually no blood accumulates. When you are aroused, the cycle reverses. Like a balloon fills with air that can't escape, your penis fills with blood that can't escape.

The corpora cavernosa (penis) was empty before, but it is now full of blood as you stand to attention. Just look at the Latin term corpora cavernosa. This translates to 'cavelike body'. Filling this cave with blood leaves less blood for the brain, so your decision-making skills could be rendered momentarily suspect. The mechanics of increasing and decreasing blood pressure in the penis is at the heart of dysfunction problems. Relaxing the arteries going into the penis is the goal of most ED drugs.

Until the early eighties, the problem was thought to be in the brain. Dr Giles Brindley changed all that by injecting his penis with phentolamine, giving him an almost instant erection. He dropped his trousers to show the world – ok, not the world, just a bunch of willy doctors (urologists)

– what he had figured out. He used the drug to relax the arteries going into his penis and BOING. The race was on for researchers to find a magic pill that could do the same thing. It turned out to be a little blue pill – the best-known drug in the world – although other medicines work as well.

Phentolamine doesn't give you a choice, but ED medications do. Also, the thought of injecting your penis with a needle isn't an option most men would ever go for. So how does the pill know to work only when you are worked up? The key is in the chemistry of muscle relaxations. Your brain controls the release of commands to make this happen. When a signal is sent to any muscle to relax the arteries, a chemical reaction starts. The chemical reaction causes an enzyme (cGMP) to react, which allows the muscle to relax. Another enzyme (PDE) works to destroy the cGMP. The PDE produced in the penis is different from other forms of PDE in your body and the little blue pill can, thankfully, stop the penis PDE for a short time while you stand proud.

The blue pill has potential side effects of blurred vision and headaches. That is because penis PDE and head PDE are similar, so some people have head problems when they take the blue pill. Another, worse, side effect is a heart attack, so read and follow the directions. Other drugs are on the market and use slightly different chemicals to achieve the same result, but they also have some side effects. Moral of this story: don't buy pills because of an e-mail. Talk to your doctor to find the cause of the issue.

Trust me, when erectile dysfunction comes to your house, you won't laugh. My motto has always been 'better living through chemistry'. We may laugh about them until we need them, but ED medications may well help us later in life.

Can corpses get erections?

While some live people struggle to get wood, corpses can sometimes get it with ease. A death erection is also known as angel lust. It has been observed in cases of hangings and strangulations. Coroners have also discovered it in corpses that have died face down. Which must be a nice turn up for the books.

The erection on a face-down corpse is easy to explain: gravity. The penis is just spongy tissue that fills with blood due to gravity if you die face down. Blood will always pool at the lowest point once the heart stops pumping.

The strangulation erection is thought to be caused by pressure on the spinal cord or cerebellum. Spinal cord injuries very often lead to a drop in blood pressure due to a loss of muscle tone in the blood vessels. This drop in pressure causes blood to pool in the extremities. For most guys, that extremity is the penis. This phenomenon was noticed at public hangings as many victims would develop erections as they swung from the noose.

Scientifically speaking

In the school I went to, they asked a kid to prove the law of gravity and he threw the teacher out of the window.

- Rodney Dangerfield

Why do burps happen?

Much like farts, burps come from extra gas that we swallow. As we eat, we not only swallow food and drink but also air and other gases. The pressure in our stomachs builds up due to this extra gas and the body burps to relieve pressure.

The valve at the top of the stomach opens like a safety valve to let some gas go. A particularly good burp will even leave a taste in your mouth as it leaves. It may be socially unacceptable in many circles, but it's better than bursting like a balloon. It is okay to burp if you are under three or over eighty-three years of age. When a baby burps, people in the room cheer. When a senior citizen burps everybody laughs.

Don't we deserve applause or laughter when we let a good one rip? Many of us can do the alphabet or sing the National Anthem in one single belch. We've spent years developing this skill and should be rewarded with adulation, not cold stares shortly followed by rancorous heartburn.

Did you know?

Scientists committed to stopping global warming are currently looking for ways to decrease the amount of methane that cows burp and fart into the atmosphere by giving them foods that create less gas.

Why do farts stink?

The old joke is that farts smell so deaf people can enjoy them too, but it goes deeper than that. The smells come mostly from aromatic sulphur compounds in the gas. Farts contain carbon dioxide, oxygen, nitrogen, hydrogen, methane and other trace gases, but the smell producer is primarily hydrogen sulphide.

Some foods create more gas than others. Broccoli, cauliflower, meat, eggs, milk and good old baked beans are all good for farting. Beans are the all-time king of fart foods, but bean farts aren't always the most aromatic. Beans contain sugars that are broken down in the intestines and gas is created. Vegetables usually cause you to parp more, but they aren't as potent. Matter of fact, much of the roughage of plant material will be broken down in the intestines and create gas. They just aren't as smelly as protein-based farts.

Cow's milk is the dark lord of fart food for many people. Cow milk contains lactose, not present in mother's milk. Lactose is broken down by the enzyme lactase and many people are actually lactose intolerant. They do not produce enough lactase and a few don't produce any at all. All of the lactose must be broken down in the colon and this breakdown produces gas... a lot of gas. Be careful! Never offer to take these people out for ice cream unless you are eating and leaving.

Is holding a sneeze dangerous?

I asked four doctors and none had ever signed a death certificate with sneezing listed as a cause of death. So I can confidently say that holding a sneeze won't kill you. Even though it won't kill you, holding a sneeze could cause you to blow your eardrums out. The extreme pressure inside your head needs to be vented somewhere. It is an urban legend that your eyes would pop out, unless it was a truly volcanic sneeze.

You can stop a sneeze with the finger-to-the-upper-lip trick. Here are other ways to stop sneezes. Don't breathe! Simple and effective, but it can lead to death if done long enough. You can also press your tongue tightly to the roof of your mouth, play with your ear lobe, or loudly say the word 'watermelon' three times. This works and gives anyone within earshot a good laugh. Force your eyelids to stay open. Your body will never let your eyeballs pop out, honest.

Of course, these methods only work with sneezes that you feel coming on. The quick, immediate sneeze is going to happen occasionally. I think you should just let all sneezes go. You will feel better and your friends will get something to laugh about. After all, laughter is the best medicine.

Did you know?

It is impossible to sneeze with your eyes open.

Do men think about sex all the time?

Depends who's asking! Blokes think about sex all the time because our urge to procreate is deeply rooted in our inner beings. It isn't our fault; it's biology. Almost all animals have an overwhelming need to continue their species. Men are no different and it's a comforting thing to blame our desires on something other than our brain. The desires are a direct result of nature's designs. Sperm is good for seventy-two hours, but the egg is only good for twenty-four, so we have to work all the time to continue the species. Even if you don't want kids, the need is contained in your DNA. So we still think about it.

I heard a long time ago that on average men think about sex every seven seconds. I don't believe that number – it seems too short. Of course, once you factor in all the teenage boys in this country, it is probably longer for the rest of us. In my older age I have been known to go at least three minutes without thinking of sex. The famous Kinsey Reports actually said that only 54 per cent of men thought about sex even once a day. They call that report the bible of sex research, but it came out in 1948! In 1948 we had very limited television (no Channel 5 for starters), no bikinis and no internet porn. But maybe they were right. So if they're right and the seven seconds is also right, some dribbling deviants are doing way more than their fair share!

What is the prostate probe?

The word digital conjures up thoughts of high-end audio and video equipment, but it has other, not so pleasant, meanings. And one of those different meanings is found in the doctor's office. When men reach a certain age, all physicals include a digital rectal exam. Digital in this case refers to digit, as in finger. And rectal means, well, rectal – there's no getting away from that. The prostate probe is a necessary evil that is really beneficial.

The doctor takes a gloved, lubricated finger and probes into your nether regions to check your prostate gland – a walnut-shaped organ that helps in ejaculation – for abnormalities. Catching these abnormalities early is beneficial because treatment done early is almost always successful. The test only takes a minute. You drop your pants and your doctor begins to probe. Just be careful. If the lights dim and the speakers start playing Marvin Gaye music, you know it's time to get out of there.

Why does hair grow out of our nose and ears?

Hair has been in those places all along. Nose hair serves as a filter for the air we breathe in. In a dust storm, we could easily close our mouths, but closing your nostrils is downright hard. So nature built in a little air filter. Ear hair also is designed to keep junk out of the ear canal. Most men notice that their head holes seem to become more full of hair as they age. We might try to pluck, trim, cut or tweeze but it just keeps growing back. You can have permanent hair removal done, but you're a man so you won't. The amount of medical research into ear and nose hair is surprisingly scant, so I'll just give you my two theories.

First theory: the adult body has a certain amount of hair follicles for its entire life. As you age, follicles migrate. Some inner working of the body causes you to lose hair on the top of your head and gravity causes it to migrate south. As it migrates south, it stops over in warm, moist places like the nose and the ears. This is similar to going to the beach when you were younger – a wet, warm place to hang out. So the nose and ears are like going to the French Riviera for hair follicles. Only the lucky follicles get to hang around at the beach, until they get attacked by a giant weed eater called a nose hair trimmer. The problem with this theory is that the hair follicles also migrate to your back, which would be more like the desert. Of course, in Britain, many coffin-dodgers move to places like Torquay, Eastbourne and the Isle of Wight. So maybe the hair follicle migration theory makes sense.

Second theory: warm, moist areas grow hair better. The nose and the ear are both, so they make fertile areas for growing hair. You can think of the ear and nose as warm, wet garden plots. Most plants grow best in warm, moist areas, so this makes sense. You can think of back hair as cacti and aloe plants. The warm-moist theory postulates that hair will always grow in the nose and ears and it just becomes noticeable as we age. It was always present, but our thick schoolboy mullets distracted people away from focusing on it. The warm, moist hair just keeps growing along. As we lose hair on our scalp, the hair growing from the nose and ears becomes noticeable. By middle age it is far more noticeable because it has less competition.

The actual reason of shifting hormone levels isn't nearly as much fun as making up random theories. Any professional research scientist reading this might want to consider writing a grant to study this offensive hair. You can start with my two theories. Until then, I'll be tweezing away on the forest of hair growing from each nostril. They'd better not join forces.

5. Technobabble

Men love technology. It's wonderful and frustrating at the same time. Using a digital video recorder (DVR) to skip tedious adverts is absolutely fabulous. High-definition television allows you to see individual blades of grass on a putt at the Masters. MP3 players with ten thousand songs are technological wonders. Best of all, I can hide my musical tastes from others. No longer can my mates see that I own all of Cliff Richard's music.

Many men love to become techno-geeks. We stare at the old VCR as it continually blinks 12:00 but will gladly spend all Saturday trying to install an eight-speaker, digitally tuned, stereo-surround system to shake the entire neighbourhood while watching movies. We love technology if it is shiny, new, expensive and takes at least a day to program.

We will dive into all the devices that let us proudly say 'I am a techno-geek'. Which is better: LCD screens or plasma screens? How do they make 3D movies? Can you use a cell phone in a metal building? How does sound surround you? And how do DVRs allow us to skip the adverts? Turn off that 84-inch flat screen and let's examine the world of technology.

How does the internet work?

The internet owes its start to the Cold War. The US Department of Defence initially funded a project in the late 1960s to provide a reliable means of communication in the event of a nuclear war. The DARPA project was designed to link the massive computers of the day. Today, the internet is a large-scale group of networked computers, routers, dedicated information lines and phone lines that speed information around the globe, using the world wide web.

You can think of the internet as a worldwide highway system. The dedicated information lines are like superhighways. These large information lines can handle trillions of cars (messages) per second. They interconnect towns, countries and even continents. The routers are exits on and off of this information superhighway to smaller roads (phone lines and cable lines) that lead to your computer. Just like your house has an address, so does your computer. This internet Protocol (IP) address allows you to send a message to another IP address and get a return response. And the best part is that the messages can be sent almost at the speed of light – 300 hundred million metres per second – right to your door.

An IP address is a major shortcut that makes the internet much easier to surf. It is really just a set of numbers, but it can be replaced with a URL (Uniform Resource Locator) that displays a name like www.suckmywebsite.com, which is much easier to remember than a string of numbers. You pay an internet service provider (ISP) for access to your local road and then you can surf for hours.

Scientifically speaking

Science is a wonderful thing, if you don't have to earn a living at it.
- Albert Einstein

How are 3D movies made?

Most three-dimensional movies use two separate cameras, each filming from a slightly different angle, but some newer ones create a second angle using computers. Couple two different pictures with a few filters and you are on your way to ducking as a dinosaur lunges at you. If you take off the glasses, the picture appears fuzzy because the images are slightly off-centre.

Each camera lens is outfitted with a polarizing lens oriented in a different direction. Polarizing filters only allow one direction of light waves to pass through. The left lens may be up-and-down polarized, while the right lens is side-to-side polarized. The key is watching it with the ever-so-cool 3D glasses.

The glasses each have a polarized lens that is aligned with the lenses used in filming. Two cameras project the film onto the screen and the polarizing filters allow only one scene to reach each eye. Your brain stacks those images up and sees in three dimensions. Many 3D movies from the 1950s used über-cool red and green glasses to filter the two images, but today most 3D glasses use polarized lenses. The glasses make great sunglasses and create a fashion statement at the same time. Don't throw out the old red-green ones either, since a few video gaming systems are rumoured to use them for newer games. Your optometrist could even put the lenses in stylish frames to improve the gaming experience. You won't play any better, but you will look better as you lose.

How does a plasma TV work?

All light is created in an atom in the same manner. Electrons circle the atom in clouds at different layers; when they are excited by heat or electricity, a few of these electrons jump to a higher level. Think about walking to the top of a water slide and when you slide down you let out a giant yell. For electrons, jumping up is like climbing that giant ladder. These excited electrons soon get bored and plunge back to their previous level. As they fall, they give off a photon of light, like your yell as you plunge.

Plasma TVs use plasma, which is electrically excited gas. Gases start neutral but can become charged as their electrons are ripped free from their host atoms. Free electrons are negative and the old host atom is now positive. The net charge overall is still neutral, but you have shifted charges. To rip away these electrons, you need extreme heat or extreme electricity. Think of a plasma display screen as thousands of tiny fluorescent lights. Each pixel or point on the screen is composed of subpixels of red, green and blue. By varying the intensity of those colours, you can create every colour except black. To create black, just dial all the subpixels to zero.

The screen uses thousands of tiny glass cells full of a mixture of xenon and neon. When the cell is excited, it emits an invisible ultraviolet photon which strikes a red, green, or blue phosphor, which emits a visible photon. Each cell is excited by electricity delivered to an address electrode for each cell. The cells are excited several times a second and the picture moves seamlessly.

How does an LCD screen work?

LCD stands for liquid-crystal display. The term liquid crystal is a bit misleading, because liquid crystals are not actually liquid but solids. However, the molecules in these crystals behave as a liquid, hence their name. Having a liquid-like appearance while actually being solid gives them the advantages of both.

LCD screens require small electric currents to manipulate the passage of light through the liquid-crystal molecules that act as a screen. The current causes the crystals to twist and let varying amounts of light through. LCD screens do not produce their own light; they just let you see existing light. Think of it like a bunch of windows opening and closing to create the picture on the screen.

The liquid-crystal display is sandwiched between polarized glass sheets, which are perpendicularly aligned to manipulate the intensity of light as it passes through the crystals. This allows it to quickly switch from displaying light images to dark images and the grey in between.

Active-matrix displays are used to enhance the picture. These displays use transistors and capacitors in a matrix, which is located on the display glass. Electrical charges are sent to particular subpixels. LCD screens create colour by using subtraction. They use filters to block out all of the colours except red, green, or blue for each subpixel. By varying the intensity on each subpixel they can create millions of colours. By actively manipulating electrical charges, you will see a sharper and clearer display on your television screen.

So which is better: LCD or plasma?

There are advantages and disadvantages to each. The technology is changing so fast that it is hard to even keep up with the changes. You can be guaranteed that six months after you buy a new technology a better one will be available.

Plasma screens are extremely bright and deliver very sharp pictures from almost any angle. They can be very big and very thin, but many have limited life spans and use a ton of energy to power each individual electrode. LCD screens are flatter, can be hooked directly to a computer and use less energy than a plasma screen, but they are often difficult to see at angles.

Advances in technology have almost made this question a toss-up. Five years ago, plasma TVs were superior, but today there's not much difference. Now it is completely a personal and budgetary choice. But don't buy a particular type on my recommendation. You really should go to the electrical superstore and buy a TV based on the recommendation of a barely literate, acne-ridden teenager.

Why will you miss your old telly?

Flat-screen televisions will be the wave of the future, but there will be one reason to miss your old cathode-ray tube (CRT). That reason is the raspberry effect. Stare at a CRT, then stick out your tongue and give it a good old raspberry. Cool, the screen will wobble. This trick won't work with your flat screen.

The raspberry effect works because your eyeballs are shaking. The screen refreshes between fifty and sixty times per second. In other words, sixty times a second the screen turns on and off. Because your eye is moving, it is in a different position as the screen is drawn. As the image of the screen traces a path across your retina, it looks like the screen is moving. You can get the same effect by blowing a raspberry at most LED displays, like the red numbers on an alarm clock or many fluorescent lights.

Next time you are at work, walk around the office blowing raspberries at all of the electronic devices. Go on – do it! Your bemused colleagues will love you for apparently making such a bold statement about your job.

Scientifically speaking

Magnetism, as you recall from physics class, is a powerful force that causes certain items to be attracted to refrigerators.

- Dave Barry

How does a DVR pause live-action television?

Digital video recorders (DVRs) are just a hard drive that your television signal is routed through. The signal is recorded into a television buffer any time the TV is on. Most of the buffers are one hour, which means that one hour of what you are watching is always being recorded. If you walk into the room fifteen minutes late, you can rewind and finish watching the movie from the start. The DVR just keeps recording the same channel. The best part is you can catch up by fast-forwarding past the bloody adverts.

Pausing live action is probably the most advertised part of the system. You can pause while you go and get some grub. When you get back, you can pick up where you left off.

Most units allow you to record one channel while watching another and a few even allow you to record two channels while watching a third. However, once the hard drive is full, you need to buy a new unit or download the information to a larger hard drive. I feel confident that the units will soon be integrated with your computer, but there is still a limited hard-drive space. Burning onto DVDs is a possible option.

One benefit of a DVR is the search feature that allows you to search by actor, programme, keywords and so on. You can record all of the great kung fu movies that run in the wee hours by recording any title that contains the word dragon. This feature is great for practical jokes. You can set a DVR to record everything containing the word Pope for an atheist friend. Or record everything starring Keanu Reeves for your friend who attends drama college. Just wipe your fingerprints off the remote if you try a prank.

How does sound surround you?

Surround sound is a given for today's techno-geek and just about any movie fanatic. Surround sound specifically refers to the process developed by Dolby, but we commonly apply it to any multichannel system. Once the purview of moviegoers, surround sound is now fairly common in home cinema setups. But getting sound to surround you has taken many years.

One of the earliest movies to use surround sound was *Fantasia* by Disney in 1940. The sound editor took separate recordings of each orchestra section and combined them in a fairly new way. He recorded tracks for each of four different speakers for the theatre. Mostly by fading one orchestra section into another, he made the classical music surround the moviegoer by moving the sound around the theatre.

Surround sound continued in smaller leaps until the 1970s. The Dolby Stereo system became the staple for theatres and this is still pretty similar to what we use today. *Star Wars* was one of the earliest movies to use this technology. By fading music from the rear to the front speakers, moviegoers felt like the ships were flying right by them. The battles seemed to come alive. The rear speakers were used for background noise and moving spaceships. The front three speakers contained all of the dialogue. In theatres, a person talking on the left of the screen will come from the left speaker and so on.

In the 1980s and 1990s surround sound came to our house and never left. The ability to simulate the movie cinema experience was amazing. And we didn't have to take out a loan to buy a litre of fizzy drink and stale popcorn. Fresh popcorn, cheap drinks, surround sound and pausing the movie while we nipped off to the loo caused home-cinema use to explode.

The earliest home cinemas came with four separate recordings: track

A, track B, the common parts of A and B and the differences in A and B. Track A went to the left speaker and track B went to the right speaker. The common parts of A and B went to the centre-channel speaker. The different parts allow the front and rear speakers to play different sounds for each side of the room. The surround decoder lowers the volume for the rear speakers and adds the desired time delay. A variety of settings allow you to mimic an arena-sized venue or a smaller theatre.

The subwoofer plays the low notes to shake the entire room. Low sounds cause your body, the chairs, your dog and your walls to shake. Sound is just air pressure shaking your walls, kind of like miniature sonic booms. Low, loud sounds have the biggest difference in air pressure. These low notes cause the room to shake and us to love action movies. Romantic comedies are fine for single-speaker television sets, but blokes want action. We want to feel the explosion. We want to duck as a light sabre slices above our head. So we pay the money and spend four or five frustrating hours to set up our home theatre. But we do it gladly, just to shake the entire house as we watch our favourite flick.

Scientifically speaking

All of physics is either impossible or trivial. It is impossible until you understand it, and then it becomes trivial.

- Ernest Rutherford

What's so great about an MP3?

MP3 players have been a staple since the late 1990s. And why not? You can fit 20,000 songs on one little player and nobody has to see that you have Boyzone's greatest hits on auto shuffle.

MP3 files are compressed audio files and there are several benefits to compressing an audio file. One, you can store more songs on a player. Two, you can download and upload them to the internet quicker. The music loses almost nothing to the untrained ear.

MP3 players can be solid state or use a computer minidrive. Solid-state players have no moving parts, so they won't skip while you are exercising or in the car. Mini hard drives are usually loaded with built-in shock absorbers to prevent skipping. At the heart of the player is a microprocessor. You direct it to load a playlist that you previously created and you are ready to play. The processor grabs the digital file from the memory and decodes it into an analogue signal to drive the speakers in the ear buds. Most cars and home stereos are now outfitted to play MP3 files, but you can, of course, hook it up to an external amplifier to hear on more powerful speakers.

No CD cases littering your passenger seat. Nobody has to see your musical taste unless you want them to. Great for now, but MP3 players will soon be replaced by something smaller and faster. Trust me, your children will eventually look at you and laugh when you say you had an MP3 player. My students laugh when I talk about vinyl records. In a few years, kids will own MP37 players that contain every song ever recorded by every artist ever and probably double as an earring or something.

How does an iPhone touchscreen work?

The Apple iPhone took the market by storm. The simple flat black box came alive with the slide of a finger, but sales took off because of the touch of two fingers. The touch screen on the new phone was better than its predecessors. The two-finger pinch to resize pictures and graphics is the epitome of techno-cool.

Most touch screens use resistive technology. Two thin glass or plastic sheets compress when touched. The screen knows where your finger is located because the compression changes the resistance of a thin metal sheet. However, the iPhone goes above and beyond and uses capacitive technology.

Tiny capacitors located behind the screen can locate your finger even if it doesn't touch the screen. By aligning an array of thousands of tiny capacitors, the phone can determine the location of two fingers and that is the secret to its success. Your body is a giant source of electric charge and capacitors transfer electrons to store that charge. One plate becomes positive due to lack of electrons and the other plate becomes negative due to the presence of extra electrons. As your fingertip approaches the screen, the plate will detect the location of your finger due to an increased capacitance.

Touch screens are now found only on high-end products, but soon they will be everywhere. Touch computer screens may replace the traditional mouse and touchpad soon. And as the technology gets better, we may even replace the traditional keyboard with a touch screen.

How do mobile phones work?

Mobile phones are just a high-tech version of the walkie-talkie my parents would never buy me as a kid. The mobile phone converts your speech into an electromagnetic radio wave by a process called superposition. Essentially, your audio wave superimposes itself on a carrier signal. The signal is sent to a cell tower that boosts the signal and sends it to the recipient. The recipient's cell phone subtracts off the carrier signal and converts the audio track back into audible sound.

In the early days of mobiles, the total number of cell areas was small and the phones were gigantic. Now phones are tiny and virtually the entire world is covered by at least one cell area. Mobile (or 'cell') phones allow us to be in contact with almost anyone in the world, which can be a good or a bad thing. So if you need to avoid a call, read the next question for a great excuse.

Can you use a mobile in a metal building?

Any building that has a predominantly metal shell will block most phone signals. The metal building acts like a Faraday cage. Michael Faraday theorized that electromagnetic, or EM, waves would travel around a hollow conducting cage. Lo and behold, he was right. The EM waves cause electrons in the metal to move in response to them and no waves can penetrate (or escape). The longer the wavelength, the easier it is to shield.

Newer mobile phones are harder to shield because they use shorter wavelengths. Another way to help phone signals is to place an external antenna that connects to an internal antenna. Even in normal buildings, the phone signal will degrade as it passes through more solid materials. That is the reason you see all of those people walking around with their phones held in the air searching for a signal. If that doesn't work, try hopping on one leg as you hold the phone skyward.

Brain fart

'Dammit, I just lost an electron.'

'Are you sure?'

'Yeah, I'm positive.'

6. Party time tips and tricks

Men enjoy a good party. We enjoy the sights, sounds and all the other distractions created by people having fun. Whether we are at a pub or a barbecue in our own back garden, we love messing about with friends. As we age, we try to drink more responsibly (or not at all) and we still have a good time. Teenage and university parties may be a distant memory, but they are probably a pleasant memory.

Most of us have a few great party stories from our youth. My favourite party words were 'Oi, watch this!' When I heard them, I knew something really outlandish was coming. I imagine more partygoers have been hurt by those words than any others. Luckily, I never said them – at least not that I remember. Now that I am older, when a friend utters those words, I make him explain what he is going to do beforehand. If it's safe, I let him go ahead.

Drunken bar tricks, six-packs and breathalyzers are all-time party staples, but the science of a party is also eye-opening. Have you ever wondered why ice sinks in whisky? Or why tequila won't freeze? Belly up to the bar and let's take a look at the science of parties.

Why does ice sink in a glass of whisky?

Ice floats in liquid water. This is the opposite of what you would expect. Solids are typically denser than their liquid counterparts and therefore sink when added to a liquid. This doesn't happen with water because as the liquid water cools to near freezing (4°C or 39°F), the water's density changes. At this point, the molecules move very slowly and are attracted to each other. However, as they group together, they bond into organized hexagonal patterns. This is the most efficient way of packaging water molecules. When water forms these hexagonal patterns, there is empty space between the molecules, which makes the frozen water less dense than the liquid form. In this case, water's solid form takes up more space (therefore less dense) than the liquid form. Thus, solid water floats in liquid water. Water is the only common substance to do this.

When we add the same ice to whisky (or almost any alcohol), it sinks to the bottom of the glass because the density of alcohol is less than the density of the ice. Barmen can even make cool layered drinks based on the differing densities of types of alcohol. This also helps because as the ice melts the water stays on the bottom and your drink doesn't get watered down. I took advantage of this when I was a wee lad by stealing some of my dad's whisky and pouring water back in. The top layer would be pure firewater and he was none the wiser. Until he finished the bottle! Luckily, he drank slowly. I was of legal age before he finished.

What is the secret to layered drinks?

Striped drinks of different colours are great conversation starters. They are carefully prepared, only to be slammed down in one gulp most of the time, but what is the secret to this alcoholic art?

The secret is density. All alcohol has a proof number, which gives the per centage of alcohol. Eighty proof has 40 per cent alcohol and 60 per cent water and some form of sugar. A general rule of thumb is the higher the proof, the lower the density. Density causes all objects to sink. Vinegar sinks in oil because the vinegar's density is greater than that of oil. A good shake will mix them up, but they will eventually settle back out.

To make a layered drink, pour the densest liquid in the shot glass first. For the next layers, a spoon will help. Turn the spoon over and gently pour the next liquid onto the back of the spoon. Continue layering until the drink is done. Now pass out the rounds and slam them away. The different tipples mix in your mouth and burn on the way down. With different coloured alcohols, it is possible to make any variety of colours. You can even make my favourite, the Buttery Nipple. But that's not something I want to elaborate on right now.

How do you crush a beer can on your forehead?

Crushing a can takes strength, speed and inertia. Inertia is the resistance to a change in motion of an object. The more massive an object is, the more inertia it has. Your head is massive (at least when compared to the can). You grab one end of the can and violently smash it into your forehead. Your head's inertia will want to keep it stationary. The can's smaller inertia will cause it to change shape. Just think about it; you wouldn't crush a barrel into your forehead. It also helps to have a Cro-Magnon forehead. Harder heads have more inertia and fewer brains.

There are far better ways to see the same concept. Take a brick and the same beer can. Place the can on the ground and drop the brick onto it. The can will crush because of inertia. The brick has more mass and wants to keep moving, so it crushes the can. This is a much smarter way to get cans ready for the recycling bin. Leave forehead can-crushing to blokes known affectionately as 'Psycho'. And never – repeat, never – confuse the brick with the beer can.

Brain fart

Most people who crush a can on their forehead regret it the next morning. Is it coincidence that most cans crushed on the forehead once contained lager? No, it's not.

How does the falling fiver trick work?

The falling fiver trick is a fun way to win a free drink. There is no better word than free. You pay nothing, zero. And you can give a little science lesson as you drink that free ale.

Here's the setup: hold a fiver by your fingertips with the note hanging down to the floor. Have your opponent place his or her thumb and forefinger around the centre of the note but not touching it. Bet him he can't catch it two out of three times as you drop it. If your opponent has been drinking, you can probably bet that he won't catch it any of the drops.

The science behind this trick is reaction time and falling bodies. As an object falls, gravity is the primary force responsible for its acceleration. All objects accelerate at 9.8 metres per second every second (that's 32 feet per second every second) near the Earth's surface (if you ignore air resistance). Since the note is perpendicular to the floor, air resistance is not much of a factor. The average human reaction time is 0.15 seconds. That is the time between when you see something and your brain can get a message to the muscles. And, of course, reaction time will be slowed if alcohol is involved.

According to physics, the note will fall over 11 cm in that same time. Half of a five pound note is about 6.7 cm in length, so someone attempting to grab it will just miss the money. Two out of three is a safe bet, since occasionally someone will time the drop correctly. The fiver falls to the floor and you win free beer.

You can also show that you have better reaction times than your opponent by dropping the note with one hand while you catch it with your other. You are just so amazing! Helps if your entire audience is comprised of gullible drunkards, naturally.

What's the secret to the upside-down shot glass?

The classic upside-down shot glass trick has been done by countless bar patrons over the years. This is also the perfect trick to amaze your six-year-old nephew. The needed supplies are readily available and the trick always works. Take a shot glass and fill it up with water. Of course, you can always use your favourite booze, but not with your six-year-old nephew. As a matter of fact, you might want to use a regular drinking glass for him. You don't want him going home and saying 'Look what Uncle John showed me with a shot glass.' You might lose serious points in the family game.

Take your full shot glass and place a playing card or takeaway menu over the open top. Any flat, rigid, nonporous surface will work. Press down slightly on the card and turn the drink over while pressing. Release the card and the liquid magically floats upside down. Make sure you line up any bets or payoffs before you turn it over. Your amazing abilities have allowed you to defy gravity.

Make sure you never try this trick in space. The trick will somewhat work, but you will eventually have round drops of alcohol floating around your spaceship. Floating around and slurping up the liquor might be fun, so maybe you should actually try this on your next junket into orbit.

The trick works because of air pressure. Air pushes on us from all sides and is actually trying to rush into the space that our body occupies. When you tip the glass over, the water is pushing down, but the air pressure is pushing up. The air pressure wins. And maybe you win too.

How do you trade whisky for water?

How about a new classic bar bet. Trading whisky for water is sure to be a hit at your next party. This trick is definitely not recommended for your six-year-old nephew. Adults only, please. Take two shot glasses. Fill one completely with whisky. Fill the other with water. Place a flat card over the shot of water, then invert it and carefully place over the whisky shot. Slowly slide the card out until you have a tiny opening between the two glasses. Sit back, watch and enjoy.

You will see the dark liquid slowly travel toward the top. It will take at least a minute for the process to finish. With practice, you can even invert the whisky shot with the card slightly off centre, but you have to get it back down on the water shot fast.

The trick isn't a trick at all; it is simple physics. Whisky has a lower density than water, so it will gradually float to the top. This trick will also work with vodka and gin. Try it. I dare you. Just let me know when it is done.

Brain fart

Trading whisky for water is just like Italian salad dressing separating in the bottle. Except Italian dressing won't impress your friends and you don't really want to be downing it in shot glasses either.

Why does helium make your voice go squeaky?

Inhaling helium is dangerous. It can kill you. NEVER suck helium from a pressurized cylinder. You can pass out within a minute and die. A single shallow breath from a balloon is okay, but no more. And make sure you breathe normally for several minutes before trying it again.

Okay, warning over! So how does a single shallow breath change your voice? It changes because of the helium (duh). But let's understand how you make sounds in the first place. Your lungs push air up the larynx and the air vibrates the vocal cords as it rushes out. By moving your mouth, tongue and lips, you create spoken words or songs. By repeating what you hear, you eventually learn how to create the sounds you want.

You even create a few sounds that you may not want. That is the reason for crazy accents; you mimic what you hear all the time. Most areas in the world have their own unique accents. Spoken or sung sounds are created by a variety of resonant frequencies from the column of air created by your larynx and mouth. The lowest frequency is called the fundamental and the higher frequencies are called overtones. The sum of all of the frequencies is the pitch and timbre (quality) that you hear.

In normal air, the fundamental frequency is the loudest. Helium is a lighter gas and allows the sound wave to travel faster. When you inhale helium, the gas moves faster and sounds from the lower frequencies decrease in loudness, replaced by an increase in the higher overtones. The overall pitch doesn't change because your vocal cords are still vibrating the same since your body is trying to reproduce the same sound. The timbre changes because the overtones are amplified. Your voice sounds like Donald Duck.

How does a breathalyzer work?

Breathalyzer is actually a brand name that is now applied to an entire line of devices. You may find it ironic that the original Breathalyzers were created by the Smith and Wesson Company responsible for most of America's handguns, although they were later sold to another company.

Breathalyzers are an easy way for the police to measure blood alcohol content (BAC) on the roadside. A urine test or a blood sample is more accurate, but most police officers don't like needles. And they definitely wouldn't want you peeing on their shoes, so the breathalyzer was invented.

Alcohol is not digested as you drink it, nor is it chemically changed in the bloodstream. As blood passes through the lungs, some of the alcohol moves across the lung membranes into the air. The concentration of the alcohol in the air you breathe out is related to the concentration of the alcohol in the blood. As alcohol is exhaled, it can be detected by the device.

Modern breathalyzers use one of three different technologies: a colour change, infrared spectroscopy, or a fuel cell. They all measure the alcohol level on your breath. Breath mints, garlic and onions may change the odour of your breath, but they won't change your BAC. Mouthwash will change your BAC, but in the wrong way because most brands contain alcohol. Gargling to fool a cop is a bad idea. Let's examine the three types from the comfort of a chair. Do not try to read and drive at the same time; it's as dangerous as drinking and driving.

The original testing method used a chemical reaction to produce a colour change. Your breath is bubbled through a solution of sulphuric acid, silver nitrate and potassium dichromate. The sulphuric acid removes the alcohol from your breath into a liquid solution. The alcohol then reacts with the potassium dichromate to form potassium sulphate, chromium sulphate, vinegar and water. The dichromate changes from a reddish brown colour to a green colour as it forms the chromium sulphate. More alcohol leads to a deeper green. The silver nitrate is just a catalyst that helps the process along and is unaffected, similar to the wingman concept when you are out at a party. To find out the BAC, a photocell shines light through the mix and compares to a control. The deeper the green, the more light that is absorbed and the bigger trouble the driver is in.

Another type of breath tester uses infrared or IR spectroscopy to find BAC. Ethyl alcohol is composed of carbon, hydrogen and oxygen. The bonded oxygen-hydrogen is what makes it an alcohol. Almost all chemical bonds bend and stretch like a spring. This bending and stretching is what the device uses to find BAC. These molecules are continuously wiggling. The wiggling changes as the molecules absorb IR waves. IR light shines through a breath sample. The absorbed wavelength of IR tells the police if ethanol is present and the amount absorbed tells the BAC.

The final type of breath tester uses fuel-cell technology. The device contains an electrolyte sandwiched between two platinum electrodes. As a breath is blown across one electrode, alcohol is oxidized to form vinegar, electrons and protons. The electrons flow from one electrode to the other through an ammeter that measures the electrical current. A greater current means a greater BAC. Protons (naked hydrogen atoms) flow to the other side and join with oxygen to form water. You may get arrested, but you are creating much-needed energy for our world.

Why doesn't alcohol freeze?

Many people keep a bottle of tequila stashed in the freezer for when the urge strikes. Frozen margaritas are a party favourite, but why doesn't tequila freeze? Well, it does freeze, just not at the temperature of your freezer. Ethanol is the alcohol in most types of spirits and in its pure state it freezes at about -114°C (-173°F). Your freezer only goes down to about -4°C (25°F). By mixing water with the alcohol, the distillers raise the freezing point, but it is still below what your home freezer reaches. Anything below 80 proof (40 per cent alcohol) may freeze in your freezer.

Don't stash a six-pack of beer in the freezer or you'll end up with busted bottles and ruptured cans. Beer and wine have a significantly lower level of alcohol content. The water in each may freeze and expand to crack the bottles. (See the next question for a cool exception to this rule.) Water is one of the only substances that expands as it freezes. Fish are glad for this fact. The ice on a lake will float to the top and allow the surface of the lake to freeze. This insulates the lake from further temperature drop. If, like most liquids, water shrank as it froze, the lake would fill up with ice from the bottom and the little fishies would die when the lake froze solid. Fish would be permanently removed from the menu in cold places.

Next time you watch a frozen FA Cup third round game in January, keep an eye on the crowd. The cameraman will find the madmen who take their shirts off as they rant and cheer. These intrepid fans probably lubricated their bloodstream with a little human anti-freeze and aren't going to die from hypothermia because their blood can no longer freeze. People will do anything to get on television.

How do you freeze a beer in thirty seconds?

One of the greatest party tricks of all time is the frozen beer trick. You can see this trick best with beer in a clear bottle, like Corona. Place a full bottle of beer in a freezer for one to two hours. Pull the beer out of the freezer and if it's still liquid, you are ready to start. Bang the bottom on a table and hand it to a friend. You can watch as the beer freezes in the bottle.

The beer will start freezing inside the top of the bottle and you can watch as the ice crystals migrate down the bottle. Within thirty seconds your beer will be completely frozen. Luckily, the beer will thaw out and be perfectly normal before long.

By placing the bottle in the freezer, you are creating a supercooled fluid. Supercooled fluids are below the freezing point of the liquid, but they won't freeze without some additional agitation. The beer is under pressure due to the carbon dioxide in the liquid. It is below the freezing point of the water in the beer, but the water doesn't freeze because there are no nucleation points around which the ice crystals can form. When you bang the bottle on the table, you release some of the carbon dioxide, creating tiny bubbles. These bubbles are nucleation points and give the ice crystals a place to start. The ice will cascade down the bottle as new ice crystals form on the previous crystals.

The trick also works with most fizzy drinks, but it will ruin the taste. You will end up with a flat bottle since the carbon dioxide is what gives it the fizz. You also end up with a flat beer, but you'll still drink it. The beer will still taste okay because the amount of carbon dioxide in the beer is less than the amount in a bottle of pop. Word of warning: sometimes the bottle will freeze in the freezer. This happens if the freezer or the beer gets bumped. Ice will form in the bottle and the bottle will break as the ice expands. But if you put the bottle in the freezer and leave it alone, you will have success.

7. Mr Fix-it

Before the birth of cable television, men would barter DIY jobs. We would trade our carpentry skills for a friend's plumbing skills. Nobody even tried to have more than one skill and our partners didn't ask (or expect) us to. Old-fashioned hardware shops were like Aladdin's caves: a giant sign saying 'No Girls Allowed' was posted on the front door. The gigantic US-style home-improvement warehouse stores were not in existence yet. Only real live professionals even knew where to find things like grout, Teflon tape and circuit breakers.

The advent of cable TV caused an abundance of do-it-yourself shows to pop up on the box. Suddenly it must be really easy to gut a bathroom, replace all the fixtures and repaint, right? It only took the bloke and his female co-host on the show thirty minutes, even with ad breaks. The hot girl on the show never even broke a sweat (I should know, I was watching closely). I wish the home-improvement shows actually showed everything that happens during the ads. Like the legions of experts that do all the hard work during the breaks and the fit girl sitting in a director's chair getting her nails done.

Anyway, the gigantic warehouse stores must have been in league with the cable companies, because they now give us a place to buy grout, Teflon tape and circuit breakers. And those stores even let girls in! Great! Now we have to change a whole kitchen worktop because the colour just doesn't go with the new paint.

Back in the day, we would just work out a way to live without something being fixed. A cracked toilet was fine as long as it didn't leak. If it did, we called a plumber or Uncle Fred, who was as good as. All walls were white because they only sold six colours of wall paint anyway. We never put a new plug on a lamp. We just avoided grabbing the cord where the bare wire showed through. Yep, the cable companies and warehouse stores are responsible for our home-improvement nightmares.

I know some people enjoy home-improvement projects and this chapter is for them. We will cover the good, the bad and the ugly of home-improvement science. I'll teach you the best ways to open a locked door, the best and worst reasons for doing projects and much, much more. Grab your tool belt as we explore the science of DIY.

How do low-flow toilets work?

Crap runs downhill, but it needs water to help it along. Low-flow toilets are designed to accomplish the feat with less water. They work for the average load, but the heavy work always demands a plunger. I personally think low-flow toilets were designed by the makers of plungers.

All toilets use water from the tank to flush into the bowl, which pushes the contents of the bowl down the drain line. The tanks on low-flow toilets hold less water (typically 6 litres or 1.3 gallons) than older toilets (13 litres or 2.8 gallons). The two types of toilets work in a similar way, but toilet manufacturers have adapted the siphons on low-flow toilets for a smoother flow. This allows these low-flow toilets to use less water. This is good for the environment, but not always good for guys. For us, the big stuff sometimes gets trapped at the bottom of the siphon and more water or a plunger is required. I am all for saving water, but using a plunger twice a day is no fun. The most reliable trick I know is holding down the handle, which allows all of the water in the tank to drain. But that won't always work for the really big jobs.

Another way to save water in your house is to install a urinal for liquid work, but I have a better idea for a new and improved toilet. Build it with a dial scale (like on a guitar amp) for each flush: 1 for easy listening all the way to 11 for head-banging heavy metal. This would save water and take care of the dirty work at the same time. Of course, the plunger companies might buy up the patent to keep this invention hidden.

One toilet manufacturing company is already onto my idea. They have a dual-flush toilet with two buttons, number 1 for liquids only and number 2 for the, well, number twos. The lower setting uses less than one gallon and the higher setting uses about 1.3 gallons. Using the numbers one and two is cute, but we need more choices. I still want a dial that goes from 1 to 11.

Now for some toilet trivia. The flush toilet was not invented by Thomas Crapper as widely assumed. Although he was a plumber of note in nineteenth-century England, the flush toilet was already in existence. He did put his company logo on all of the toilets he sold. Some people assume that crapper became slang in the United States for a toilet after GIs returned after seeing the Crapper name on so many English bogs. This actually may be complete crap, since the nickname was already used in England before that.

Brain fart

Low-flow toilets were definitely not designed with the average bloke's bowel movements in mind.

How do you pick a lock?

Picking locks is another skill learned in movie action-hero school (along with hot-wiring cars and making gun silencers). I can tell you where I learned this, but then I'll have to kill you. It is an easy concept to understand but incredibly difficult to master.

We'll start with the basic pin-and-tumbler style lock, the most common lock in the world. The lock has a centre core that connects to the bolt. The centre core turns freely when the correct key is inserted, or when an expert picker is hard at work. The centre core has three to seven pairs of pins that extend into the lock housing. The pin pairs rest atop each other and each pair is held in place by tiny springs. The pins are different lengths corresponding to the ridges on your key. When the correct key is inserted, the bottom pin of the pair will just clear the cylinder and will rotate freely to open the lock.

Picking a lock takes at least two tools: a tension wrench and a lock pick. The tension wrench is just a thin L-shaped piece of metal designed to turn the core. The pick looks like a dentist's tool, long and thin with a tiny hook on the end. Insert the wrench, twist the lock slightly and maintain this tension. Slide the pick above the wrench and locate each pin. Attack the tightest pin first. Using the pick, press up on the pin until you feel it pop free. The pin will actually rest on the top of the slightly turned core. You should be able to twist the tension wrench a little more with each pin. Repeat until you have done all of the pins. The bolt will open.

You can pick a lock with a paper clip and a screwdriver, but only if you are a real-life action hero. Most professional burglars have a set of lock picks, which are illegal to carry in most areas. Just be aware: picking someone else's lock without their permission is illegal.

Can you open a door with a credit card?

I'm not talking paying for a hotel room here. This skill is also taught at movie action-hero school, but many of the students already know this before the class. I had an entire school physics class that could open my classroom door with their student ID cards. One student taught everybody else how to do it. When I asked who opened it, they each pointed at the person sitting to their right. Luckily they were good kids. I hope criminals never figure out how to do this. I've said it before and I'll say it again: make sure this book doesn't fall into the hands of anyone of questionable morals!

Be aware that opening a door with a credit card only works with certain locks. This method will not open a dead bolt. Also, this method can ruin credit cards. It's better to try this out with the fake ones that come with unwanted credit card offers. Push the door in as far as possible. Slide the card between the frame and the door knob directly in line with the bolt. Bend the card till it almost touches the door knob. Now bend the card toward the frame and open the door. This will work on many doors due to their design. The bolt is bevelled, which allows you to close the door with the bolt out. The bevel helps the bolt recede into the door. By sliding in your credit card, you're actually pushing on the front of the bevel and pushing the entire bolt back.

Again, entering someone else's house is illegal unless you have permission and some locks have an extra metal plate installed to foil this method. If the lock is backwards, you can cut a U shape in your credit card and pull the card to open it. By the way, if your doors open this easily, call a locksmith and get them fixed. Science has shown you at least two ways to get inside when you're locked outside. But there is a third way.

What's the best way to kick in a door?

Kicking in a door should only be done in an emergency setting and only on your own door. There are a few rules that you must understand before you try to kick in a door.

1. Breaking and entering is illegal.
2. Try the door knob first and look for open windows.
3. If you have time, try picking the lock as previously shown.
4. Never try to kick in a metal door.
5. Always kick in the direction the door opens.
6. Call a locksmith, unless it is an emergency (or the director says 'Action').
7. Always refer to rule 1.

Television cops make kicking in a door look so easy. And it is for them. They are actually sitting in a chair while a stunt double kicks in a balsa wood door. Most of us don't have a stunt double. We have to do our own door-kicking.

The science of kicking a door in is fairly simple. Apply enough force until the doorjamb breaks. You can use a sidekick or a front kick, but always have your heel strike first. Kick through the door by aiming for a spot 30 centimetres on the other side of the door. Just like in golf, a good follow-through is essential. Aim right below the door knob or deadbolt. If it has a deadbolt, keep kicking. Eventually you will break the doorjamb and the door will open. Or you will break your foot and stop trying.

How does super glue work?

Super glue is truly super, at least compared with white glue. It will glue most substances in a manner of seconds, including your fingers. It will actually glue fingers together much better than it will many of the things you're trying to fix. Super glue is not much different from the two-part epoxy previously described, which uses a chemical reaction to form a permanent adhesive. You mix the two parts together, wait a few minutes and a permanent bond is created.

The number-one ingredient in super glue is cyanoacrylate, which is an acrylic resin. All cyanoacrylate needs to harden is water and almost every surface has microscopic droplets of water on it that help speed the process along. Wiping the broken piece with a damp rag will make the process even faster, but adding more super glue will slow it down, because more water is needed to use all of the cyanoacrylate molecules. Come to think of it, using super glue in the Sahara may not work so well. Anyway, after wiping down the broken piece, just put a little glue on the broken piece and hold together for five seconds to get a fast, permanent bond.

Fingers are the perfect gluing surface for super glue. They are rough and have plenty of places to create strong bonds. Also, your body is full of water, so the cyanoacrylate can sink in and harden. Faster than you can say 'oh bother' your fingers are glued. Nail polish remover will help to dissolve the bonds.

Super glue is even finding its way into the hospital to close wounds. Normal super glue contains a chemical that kills skin cells, but medical super glue replaces that chemical. If they ever sell that over the counter, we will be able to take care of our own power toy (oops... power tool) accidents.

How do portable nail guns work?

Nail guns come in two basic varieties: compressor driven and fuel-cell driven. They both accomplish the same task, but the fuel-cell types have a few advantages. Cordless nail guns are relatively quiet, except when they fire. Compressors are noisy. I'm talking front-row-heavy-metal-concert noisy. And the compressor nail gun is also connected by an air hose for you to trip over. Cordless nail guns give you the freedom to roam.

Cordless nail guns use a fuel cell, most often filled with liquefied petroleum gas and a spark plug powered by an internal battery. An internal piston, similar to the ones in your car engine, pushes the nail into the wood when the nail gun safety trigger and trigger are depressed simultaneously. The fuel cells can get expensive, but that is the cost of freedom.

The nail gun fuel cells are also different from the fuel-cell technology that is always mentioned as the future of automobiles. Nail gun fuel cells are just containers of fuel, but car fuel cells are devices that deliver electricity from hydrogen and oxygen. This process is just a reverse of one of your high school chemistry experiments. Passing electricity through water causes hydrogen and oxygen to be released. Car fuel cells just do the opposite.

Owning a portable nail gun will cause you to attempt home-improvement projects you never would have tackled before. Crown moulding used to be for experts, but with a nail gun and a compound mitre saw you can look like a pro. But remember: nail guns are proper dangerous! A safety trigger must be depressed along with the trigger to fire the nail. It is possible to depress the safety with your foot and nail your foot to the floor. Don't do that, there's a good lad.

How do gun silencers work?

Professional hitmen have used them for years in their quest for the kill and many of us 'normal' people pretend to use them in our computer games. The sound from a gunshot is created by the exploding powder as it ignites to fire the bullet. Sound waves are vibrations transmitted through a medium, such as air, and the speed at which the wave travels depends on the properties of the medium (temperature and pressure when talking about air). There is a tremendous amount of air vibration that comes with the explosion of a gun firing and if you can reduce one of the properties of the medium (temperature and pressure), you can reduce the sound wave.

Silencers work similarly to the silencer on your car. They both increase the volume at the end of a barrel. The exhaust gas now has a place to expand and cool before it finally leaves the silencer. As the gas expands, the pressure drops and with the dropped pressure comes reduced noise. The silencer also contains baffles – small metal plates inside the silencer that help dissipate the sound energy. The bullet leaves with a little 'pffft!' instead of a large bang. Most silencers also slow down the bullet to subsonic speeds so there isn't a little sonic boom.

8. The call of the wild

Animals bring us tons of enjoyment and many laughs. Companionship and a sympathetic ear are all great reasons for having a pet. Your dog never tells you to stop whining when you rant about a bad ref during a game. Your cat just sits and purrs as you lament a lost job or lost love. We love going to the zoo, so our love of animals extends well beyond just pets. Even the largest theme park in the world is designed around a rodent.

We also fear certain animals. All guys are allowed one

irrational fear of animals. You are free to hate spiders, snakes, cockroaches, or even vicious attacking rabbits with long, sharp teeth.

In this chapter we'll examine our love for animals and a few animal myths. Can you outrun an alligator? Why is man's best friend a dog? Do men inherently mistrust cats? Do bulls really hate red?

In addition to animals, we love all parts of nature. We gaze at a beautiful sunset with the one we love. As we stare, we wonder why the sky appears red at dusk but blue during the day. We love diamonds, gold and silver. What makes some stones precious? What makes light sweet crude oil so expensive? Enquiring minds want to know! So grab your dog lead and let's examine the world of animals and nature!

Why do men love dogs?

Dogs long ago earned the moniker of man's best friend; it is a badge of honour for most pooches. Your dog greets you at the door and licks your face, regardless of what mood you come home in. Your dog is the perfect conversation starter on a Saturday trip to the park. Almost no woman can pass by a cute dog without saying something. Your dog will also bark to wake you during a fire or burglary.

Loyal, friendly, forgiving, great listeners and icebreakers with potential dates – dogs enjoy the company of man. If you scold a dog, it will sulk into the corner only to beg forgiveness a few minutes later. Dogs don't need flowers, chocolate or an expensive dinner to forgive you when you take a bad day out on them.

Of course, like in any relationship, dogs often make you mad. They occasionally leave little dog bombs of recycled food in your house and occasionally want to hump the leg of a guest. Dog humping is really funny if it happens to an annoying relative, but it's not so funny if it happens to your rich old auntie. She might just cut you and the dog out of the will. It's also definitely not funny if a 'special friend' is paying the first visit to your house.

Dogs are man's best friend because of their loyalty – a trait that ranks number one on most men's lists of desirable traits. We like our human friends to be loyal. We like our sports stars loyal. We hear them say 'It's not about the money' and then we get angry when they sign a huge contract and leave our favourite team. And most of all, we like our pets loyal. Dogs win the all-time loyalty award. They stick by us no matter what.

The science of why we love man's best friend is interesting. Our brains have interesting little chemicals called endorphins. Endorphins are a type of hormone that reduces pain and produces a calmness that makes us feel good. This hormone actually acts in much the same manner as morphine and it is completely legal. Endorphins cause the great feeling we get after a run or an intense workout, but more subtle things can also cause this reaction. A good laugh and getting outside are two subtle ways to increase your endorphins. Dogs force us to walk them whether we want to or not. And of course, we always get a good laugh when the dog lets one rip. Dogs help us smile and smiles lead to endorphins. Don't you just love science!

Why are dog farts silent?

Dogs bring new meaning to the term SBD [silent but deadly]. They pass wind, walk to the other side of the room and leave us to gasp in terror from the smell. Dogs create gas in their bodies the same way us humans do: by what they eat and how much air they gulp down. But why are dog farts mostly silent? Here is my theory on dog farts, backed up with explanations. Feel free to research this topic if you need a dissertation topic.

First, dogs don't have arse cheeks – therefore they don't have much to vibrate as the offending gas leaves. All sound is created by vibrations. Without cheeks, the gas just escapes the opening with a minimum of vibrations created in the air. This would be like playing a clarinet without the reed. You could still get air through it, but it wouldn't create much sound.

Second, dog sphincters are looser than human sphincters [so I'm told]. This is similar to the balloon-neck trick we used to annoy anyone within earshot. If you stretch the neck tight as the balloon empties, you get a high-pitched shriek. But if you just let the balloon go, it will fly off in a mostly silent rush of air. Of course, this leads to the question: why doesn't my dog fly across the room after letting one rip? I think that has to do with the weight of the dog. Maybe little rats-on-a-rope do take off after a particularly gaseous exchange. Of course, a few dogs do make noise as they float an air biscuit, but most just sneak one in and leave us choking.

Can you outrun an alligator?

It appears that no credible research group has ever measured the speed of an alligator, but you have to love the Australians. They have measured the speed of Australian freshwater crocodiles and for me that is close enough. Australian freshwater crocodiles top out at a whopping 11 miles per hour. That's faster than I jog but slower than I run while being chased by an angry twelve-foot reptile with giant teeth.

The science of outrunning gators comes down to speed versus acceleration. Gators are extremely strong animals that rely on their strength to survive. They are lurkers who lay hidden in the water and wait for unsuspecting prey to wander by. But don't worry, if you see a gator, you can outrun it.

Gators are not fast, but they are quick. They rely on the element of surprise to capture food. They can lunge one body length in a blur of jaws and teeth. Over that one body length, they have incredible acceleration. They can reach full speed in one step. However, they slow down as fast as they started. If you can avoid that lunge, you are safe and the gator will go hungry. Running in a zigzag pattern will give your friends a good laugh, but it won't necessarily help you. And trust me, if a gator makes a surprise lunge at you, your brain won't think about zigzagging.

Even so, it is probably a good idea not to walk right by the edge of a crocodile-infested river. And definitely don't limp right by the edge of a gator-infested river. Gators tend to feast on already dead carcasses and injured or weak animals, so as long as you're not already dead but are in relatively good shape, avoiding being reptile food should be relatively easy.

Could cockroaches survive a nuclear blast?

Cockroaches are resilient. You spray them with bug spray and they just laugh at you as they scurry off. You step on them and they manage to find the tread of your shoe and limp off to have more babies to infest your house. They are bloody hard to get rid of. Of course, a nuclear bomb would get rid of them. Or would it?

Any roaches close to ground zero would be vaporized, so they don't always survive nuclear blasts. But cockroaches further away from ground zero may have a better chance. They may actually have a better chance than us people.

To understand why, we have to look at radiation tolerance. A rem is a measured dose of radiation that causes a measured amount of human tissue damage. A dose of 800 rems or more is considered lethal for humans. The standard American cockroach can withstand up to about 67,500 rems before it dies. For some reason, German cockroaches are even tougher; they can survive over 100,000 rems. Cockroaches can also bury themselves in the dirt for months at a time, which can help them avoid fallout.

Cockroaches are tough. The flip side of that is humans are wimpy. Only 800 rems and we are toast. We need to get tougher as a species. Maybe we should forgo the lead sheet at our next dental X-ray to help toughen us up. Maybe our dentist could X-ray each tooth individually. A single X-ray is only in the region of one-tenth of a rem, so you would need plenty. It might be easier to just get rid of all nuclear weapons, which gets my vote. Cockroaches and people would both be safer.

In my own scientific research, I have discovered that smooth-soled dress shoes are more deadly than nuclear bombs. My dress shoes have killed more cockroaches than I can count. Trainers give the roaches a chance. Each roach that limps away can produce up to 400 eggs in its lifetime. That doesn't sound bad until you imagine 400 roaches in your kitchen at the same time. And also, all 400 babies can each have 400 more. And if they do, invest in more shoes and leave the nukes alone.

Did you know?

A cockroach can live for over a week without its head before it starves to death. Charming.

Do men distrust cats?

Men fall into two distinct categories: cat lovers and cat haters. No middle ground here. You either love them or you hate them. I think this stems psychologically from our childhood. If you had cats in your house before the age of ten, you will most likely be a cat lover. No cat in the house as a child and you will be a hater. The science of memories is really quite amazing. Episodic memories are related to single episodes. These memories are a collection of what happened as we aged. We store not only the actual memory but also any pleasant associations. If we had a cat in the house as a kid, we usually have great episodic memories that contained the cat. If your first encounter with a cat was lost amid an allergic reaction or an unprovoked scratching, the episodic memory will not be good.

What about people who just hate (or love) cats and never had one growing up? That's due to semantic memory. Semantic memory is concept based and unrelated to any specific experience, which means you make a value judgement (cats are inherently evil/good) without any experience to back it up.

Cats and dogs are portrayed as mortal enemies. So if a dog is a man's best friend, then by natural extension that means men are going to distrust cats. Cats and dogs have peacefully coexisted in many houses, but the cat is always in charge. Cats can also be loyal, but it has to be their idea. Cats can also be trained not to leave stinky piles of crap around the house and if they do the dog will clean it up. There's one benefit of having both cats and dogs in your house: you won't have to clean the litter tray as often. Talk about the ultimate recycling setup.

Cats are the antithesis of dogs. Where a dog will spend three hours playing Frisbee with you, a cat would rather take a nap. Dogs greet you after a hard day. Cats might, if they haven't had a hard day. How does a housecat ever have a bad day? Alley cats have to fight for life and food, but a housecat gets three squares and tons of places to sleep. Dogs bring you the newspaper; cats bring you dead rodents.

If you ever watched Walt Disney movies as a child, you will also develop a distrust of cats. This distrust is related to what we learned over the years. If we continuously see cats portrayed as evil, we eventually assume all cats are evil. In most Disney movies, cats are portrayed as sinister creatures. They sneak around always getting man's best friend in trouble. The classic anti-cat movie is *Lady and the Tramp*, but many others show the same sneaky, troublemaking side of cats. Walt Disney truly hated cats. Of course, Walt's empire is built around a mouse, so his hatred of cats is natural. I wonder if he was born that way or if he had negative experiences with cats growing up?

I love cats and they elevate my endorphins, but I have good cats. If you have an evil, sinister, Disney-esque cat, you won't have an endorphin release. If your cat is a loner, maybe you should try putting your cat on a lead and taking it out in the sunshine for a walk.

Did you know?

Cat urine glows under ultraviolet light.

Do sharks really have two penises?

A trip to the aquarium is great fun for the family, but it can lead to some embarrassing questions if you go with curious kids. The tunnel through the shark tank is often the scene of my favourite 'what-is-that?' moment. Most male sharks have two long appendages hanging out their underside near the rear. These gigantic appendages are called claspers – an easier word to use with kids – and in reality these function like a shark penis. Male sharks are very proud.

Why two? Shark experts think it allows them to mate on either side. Swimming and having sex can't be easy, so two penises increase the chances of a successful copulation. Most of the time only one clasper is in use, but there have been a few reports of some sharks using both at the same time. What a lucky lady!

Brain fart

You may have noticed that sharks don't have arms, so male sharks will bite to hold on to their lady friends. Next time you see a shark with scars, just think – it could have been from a night of fishy passion.

Do bulls really hate the colour red?

No, bulls are actually colour blind. Bullfighters still use red capes because of tradition. The colour red probably symbolizes blood. By the way, the bullfight never ends up good for the bull. You give up an afternoon to watch the bullfights; the bull, unfortunately, gives up a little more.

The human eye contains two separate receptors: rods and cones. Rods determine the intensity of light in black and white. Cones allow us to see colours. Cows' eyes don't contain cones. The ability to perceive colour actually takes a much larger brain than the bull has. Our brain is large enough to see colour: the bull's is not. Bull's brains are large enough to see female cows, though. And the bulls keep the herd growing so us big-brained people can turn them into T-bones and leather couches.

The fact is that many animals see exclusively in black and white. Some animals have limited colour blindness. A human eye contains three different receptors for the three colours of light: red, green and blue. If a particular animal species is missing one or more receptor, the animal is colour blind. Dogs and cats are thought to be able to see some colour but not the entire spectrum. They also have more rods to help with their night vision. Daytime birds have a full range of colour vision. That is why male birds are so brightly coloured – to attract the ladies. Owls have virtually no colour vision.

Whether you are a matador or someone running with the bulls in Pamplona, it's movement and teasing that bulls hate. If you stand in front of a bull and wave a green tablecloth at him long enough, he'll get pissed off and charge. If he gets mad and you keep teasing, he'll charge again.

9. The fairer sex

Finally a chapter containing information about women from a bloke's perspective! Nothing excites and confuses men more than women. This chapter will attempt to explain the science of the female body, mind and soul. PMT, women's intuition, the 'extra rib' and more scary womanly things will be covered. You may know how a fuel injector works, but women will always be a mystery to most men.

We won't always admit it, but we spend many hours

wondering how women work. We admire women. We question other blokes about women. We even occasionally ask women themselves, but we still don't understand them. However, in my years of studying women, I have come to a few scientific 'truths' that I will share with you.

You must not let women know about the existence of the following few pages. Read them, but move the bookmark when you are done. You know that she will peek at this book to see what you are reading if it's left out. (Have you ever peeked inside a woman's book to see what she was reading? I didn't think so.) Only men will understand these truths I'm about to share. You are allowed to discuss the ideas with other men but not with your better half under any circumstances. So turn off that chick flick as we examine the world of women.

What is 'women's intuition'?

Not to be confused with the Women's Institute (the WI – they of *Calendar Girls* fame), 'women's intuition' is defined as the female art of being able to perceive something without conscious reasoning. Women's intuition has been around since the dawn of time. Think your mother had eyes in the back of her head? Nope, it was her woman's intuition. All guys understand Spidey Sense and we are okay with it. Women's intuition is just Spidey Sense without the spider bite. Maybe Peter Parker was just in touch with his feminine side. It works for him. At last count it has saved his life over a thousand times.

Men have been taught cause and effect for years in science classes, sports and life. If we do something, it will cause something else to happen. If we shank the ball past an open goal in a football match, we may not get the ball again. Cause and effect doesn't allow most of us to act on a gut feeling. We want proof. Blokes just need to reread all of the Spider-Man comics and watch the movies to get better in tune with our intuition.

Many experts claim that women's intuition is due to the fact that women are better at reading nonverbal communication and this may be true. Women can ferret out exactly what is meant by the slightest smile or gesture, while men are simple breeds who only notice things when we are knocked over the head. Although a study found men were just as adept at picking out fake smiles as women, men still struggle with acting on that fact. Pretty people could lie to us all day and we would gladly believe them.

Men have always been encouraged to make logical conclusions. We have been trained to act logically by parents, teachers and so on our whole life. Logical means we have to think and act rationally. We are taught that everything follows a set progression. The problem is that relationships aren't always rational. We need to learn to be irrational by developing our intuition.

Some experts think that women's intuition works because women are more in tune with their feelings. From almost day one, girls have been encouraged to act on their feelings. They are allowed to say 'women's intuition' and get an immediate out. Meanwhile, men were taught to hide their feelings and be tough. Those lines are starting to be blurred nowadays, but they are still there. Whatever the reason, most women have it and most men don't.

Scientifically speaking

Intuition comes very close to clairvoyance; it appears to be the extrasensory perception of reality.

- Alex Carrel

What is PMT?

PMT stands for premenstrual tension, which leads me to wonder why it's not PT. Maybe PT was already spoken for by the physical exercising crowd. Many men would probably (quietly) say PMT should be a four-letter word. PMT is real and affects women to a variety of different degrees. It is a collection of up to 150 symptoms that may all appear at the same time. Depression, mood swings, headaches, fatigue and insomnia are just a few of the symptoms. PMT starts about two weeks prior to the start of menstruation and tapers off as menstruation is reached. Medical experts think it's due to changing hormone levels, but the truth is they don't know for certain. The numbers vary from different sources, but 60 to 85 per cent of all women who are menstruating experience some form of PMT. The level can vary from 'I'm cramping' to 'Drop the remote or you die!' Some doctors even say it is a socially constructed disease and not a real malady. I feel confident that all of those doctors are male and clueless and currently in hiding.

I am one of the lucky ones; my partner has never made me give up the remote. But growing up with a mum and a sister, I can give some advice for dealing with PMT: kill it with kindness. When PMT is camped at your doorstop, you don't want to debate paint colour, rental choices at Blockbuster's, or dare to come home late. The most important thing to remember about PMT is she's right and you're wrong. The words 'Yes, dear' have also been known to be effective up to a point.

Do women have extra ribs?

A long-standing myth is that women have one extra rib. Most normal skeletons – male or female – have twelve sets of ribs, no more, no less. The extra rib story comes from the Bible, which states that God removed a rib from Adam so he could have a partner. I am confident Adam was glad to have a woman around, but I wonder if he would have willingly given up a rib. Maybe Adam started out with a cervical rib, or maybe he only had eleven after surgery. Either way, most normal skeletons have twelve now.

There is a rare genetic disorder that can cause people to be born with an extra rib, called a cervical rib. This extra rib grows out of the lowest cervical (neck) vertebra. It is usually found only on one side and affects less than 0.5 per cent of the population. This extra rib can lead to neck, arm and shoulder pain since it presses on the nerves on that side of the body. In a few rarer cases, people actually have one on both sides.

Did you know?

Babies are born with kneecaps but they won't show up on an X-ray. The patella in babies is actually cartilage and doesn't start to ossify (turn to bone) until the child is three to five years of age.

Can males get pregnant?

In the human world, no. But in the seahorse world, men are the ones who get pregnant. Seahorses are one of the most curious of all fish because they have a horse head, a monkey tail and the males get up the duff.

As with most species of animals, the male shows off to attract a mate. Once a suitable mate is found, the male starts the 'dance of love'. The couple then intertwine their tails and begins a long, slow courtship. After up to eight hours of cuddling, the female deposits her eggs into the lucky guy. Male seahorses have a pouch that can contain all of the baby seahorses as they grow. During this time, the male moves very little and loses all of his colour (and not even a sofa and a TV in sight). Eventually the male undergoes a painful looking birth as he releases approximately two hundred baby 'fry' into the water. Dad never gives them any help after that, but neither does the mother. Only a few of these babies reach adulthood to repeat the dance of love.

There is a recent case of a transgender female to male who became the first human bloke to get pregnant. Although pregnancy was possible for him (her), it isn't possible for biologically born males. Sadly, some men seem to try to compensate by acquiring massive beer-guts.

Did you know?

An elephant's pregnancy lasts 22 months. Probably not the science fact to mention to your partner if she complains about being pregnant.

How can you tell if she hates your gift?

You'll know. Up to 90 per cent of all communication is non-verbal and her body language will tell you if she is upset with her gift. Her eyes sag a tiny bit, even if she is trying to spare your feelings. Her smile involves only the mouth – the sign of a plastic, fake smile. True smiles light up the entire face. To see a true smile, surprise her with a diamond tennis bracelet for absolutely no reason. To see a fake smile, give her a vacuum cleaner and a set of new tea towels on her fortieth birthday at a surprise party with your best friends. Psychology can be a wonderful branch of science, but it won't help you if you bought her a vacuum cleaner for a gift. What's the matter with you, anyway?

The problem is that men don't read body language well. Committed blokes learn to read their partner's body because of the time they spend together and the trial-and-error method. You screw up a trial and learn a little more about what each gesture means each time. We just need to learn to apply that body language intelligence beyond the house. Learning to read our colleagues would be a great skill. Salespeople either become very good at this or they find a new profession.

Body language may be the bulk of it, but blokes prefer the spoken word. The spoken word can usually tell you if she likes your present or not, but you have to notice the body language to be sure. If she opens the gift and repeats what it is, she hates it and you are in trouble.

There is an exception: if she repeats what the gift is while beaming a hundred-watt smile and jumping into your arms, you're safe. Jumping into your arms and planting a sexy kiss on you is not subtle. Remember, boys don't like subtlety. But we do like sexy kisses and hundred-watt smiles.

10. Food science

Most normal men can cook; it just takes a chunk of meat and an open flame. And a stainless steel gas grill, a titanium barbecue tool set and a comedy apron that says 'Danger, Men Cooking' to complete the meal. Many men are grilling snobs. They can debate the advantages of charcoal briquettes, disposable packs, marinades and dry rubs for hours. Whether you are a snob or just an ordinary griller, the people you feed are going to be happy. Food just tastes better when it has been barbecued to a delicate shade of black over an open fire.

Even beyond grilling, most of us are fascinated by food. I have a close female friend who says she would prefer to take a pill to eating. You will never hear a man say that, never. We love to eat and we like to joke about food, with fruitcakes taking special abuse. We probably all have a favourite food cooked by our mums, wives or girlfriends (or supplied by a local kebab or curry house). We watch in fascination as a determined little guy wolfs down sixty hot dogs, but few of us appreciate the science behind food. So turn up the flame as we examine the wonderful world of food science.

Why is there metal in your microwave?

Many of us have started a fire by accidentally microwaving a piece of metal. And maybe even a few of us started a fire on purpose using a piece of metal. Not me, Mum! Big boys did it and ran away! Metal pieces start fires, so how come my oven has a metal rack? And come to think of it, doesn't my microwave oven have a metal shell?

The metal shell of the microwave oven actually helps the cooking process along and the walls reflect the microwaves. The electromagnetic microwave causes free charges in the metal wall to accelerate and absorb the original microwaves. As these charges accelerate, they emit new microwaves back into the oven. So the waves essentially bounce off the walls back to the food. Even the metal screen covering the glass window does the same thing. The openings of the screen are so small the microwaves can't get through and are reflected back inside. Typical microwaves have a wavelength of twelve centimetres and each screen opening is only a few millimetres.

The problem with metal in the microwave oven comes from sharp points like those on aluminium foil and twist ties. If enough of the free charges pile up at points on a metal surface, they will jump off into the air. Spark!

This spark can then ignite something else and you have fire. Never put sharp metal objects in a microwave, unless you want a light show.

What causes beer streamers?

Beer is a lovely liquid that gives many men the ability to dance and even talk to female strangers, or at least try. Beer also offers several lessons in science. Pour beer into a glass and watch the bubbles form. The bubbles in turn form beer streamers, tiny rising rivers of bubbles that originate from a few points inside the glass. Their mesmerizing dance as they rise to the surface captures our attention. What causes them and why do they rise?

The bubbles are formed when carbon dioxide molecules begin to form invisible microbubbles at imperfections (or possibly at dirt particles inside the glass) on the inside wall of the glass, called nucleation points. Once enough of these microbubbles join forces, they begin their hypnotic rise to the surface. The bubbles actually grow in size as they rise since the pressure from the liquid decreases and will cease when the beer becomes flat. And, of course, the bubbles rise because they are gas, which is less dense than the liquid. The same observation can be seen in champagne and in clear fizzy stuff like lemonade. The formation of these bubbles is similar to clouds forming by water vapour condensing around dust particles.

An interesting note: beer bubbles rise slower than champagne bubbles. Leonardo da Vinci (1452–1519) first explored this concept as he studied bubbles rising in various liquids. The study of beer bubbles would be a perfectly reasonable postgraduate study topic. I mean if Leonardo could get away with it...

What causes ice cream headaches (brain freeze)?

Picture this: a hot summer's day and a tasty '99' ice cream [now retailing at £2.50, but that's inflation for you]. An unbeatable combination. To combat the heat, you take a large lick or bite. It tastes good, but here comes the pain – a stabbing pain that causes temporary insanity. You want to scream as your forehead throbs, but then it disappears as fast as it showed up. So you go right back to the cone for more.

Most of us know this experience as a brain freeze and the pain is excruciating. But is your brain really freezing? No, it is just fooled into being cold. Run your tongue along the roof of your mouth and you'll feel a bump. The area behind that bump is responsible for the brain freeze. Directly above this area is a collection of nerves that feed into your brain.

Eating ice cream or drinking a Slush Puppie too fast causes these nerves to get cold. Your brain dilates the blood vessels to deliver extra warm blood to 'thaw out' your brain. This extra blood flow leads to the debilitating pain you experience. It goes away after twenty to thirty seconds, but it can bring you to your knees while present.

Ice cream headaches only occur in about one-third of the population, so some people are lucky. I suffer and will continue to do so. You can get rid of brain freeze by pressing your tongue against the roof of your mouth. Your tongue warms up the nerves and slows down the rush of blood. You can completely avoid ice cream headaches by eating more slowly. Letting the ice cream melt more in the front of your mouth also works. You could also just avoid eating ice cream altogether... nope, didn't think so.

Can beer batteries save the world's oil supply?

The beer battery, or getting electricity from beer, sounds like a great concept. Of course, many people would argue that beer is more important than electricity. However, you will be pleased to find out that beer batteries don't waste any beer. They are just a catchy name for a microbial fuel cell (MFC). Beer battery became the accepted name after Foster's Group brewery and Queensland University combined forces to work on MFCs, which use the wastewater left from the beer-making process to create electricity. You get beer and electricity!

MFCs use bacteria to break down the alcohol, sugar and starches that are left over from the brewing process. After the bacteria goes to work, you are left with electricity, clean water and carbon dioxide. This process will work in most food and beverage creation processes. It's a great way to clean the wastewater and you get a little electricity to boot.

The world definitely needs to do more research into MFCs. I feel confident many people would be willing to help drink the beer. More beer equals more wastewater and more electricity. So lift a pint to help the world. If we could just get science to work on the urine battery, then beer could possibly save the world's oil supply in the brewery and the bathroom.

Brain fart

Support bacteria – it's the only culture some people have.

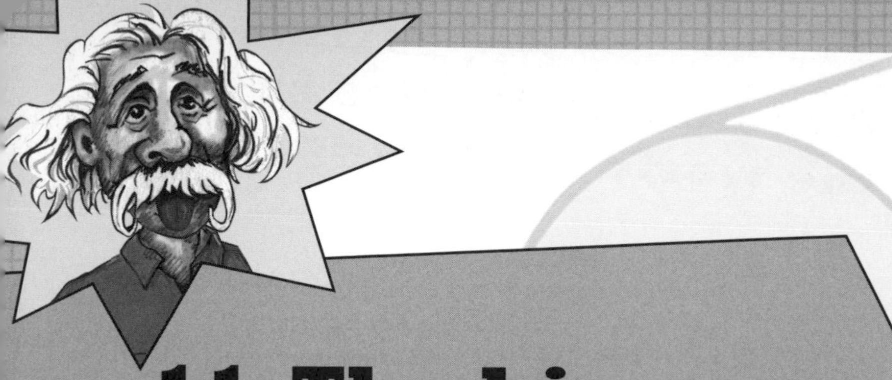

11. The big ones

The manliest things in the universe are duct tape, silicon spray (WD-40) and the remote control. In the manliness pantheon, these three sit at the right hand of fire. For men, fire is number one, but these three fall closely behind. Man-inventions invented for men by men; it just doesn't get any manlier.

Duct tape and silicon spray allow you to fix anything. If something is stuck and you want it unstuck, just use silicon spray. If it isn't stuck and you want it stuck, just use duct tape. These two inventions allow any man to improvise on the fly. No job is too tough with these two in your arsenal.

The remote control is definitely a bloke thing. In the old days, a remote wasn't needed. You had three channels and a knob. With the advent of cable and satellite TV, a remote is essential. No longer do we need to watch adverts or anything that bores us beyond our six-second attention spans. Two hundred channels and you still may find nothing to watch, but you get to enjoy the ride. In the dark ages of three channels, we always wondered if we were missing something good on another channel. With a remote control and two hundred channels, we know that we won't miss anything.

Let's examine the science behind the big three. How do remote controls work? How many remotes does a bloke need to feel manly? Why does duct tape rule the adhesive market? How did WD-40 get its name? What happens if you spray duct tape with silicon spray? Grab a seat in your easy chair as we examine the manliest three inventions on the planet.

How do IR remote controls work?

Throughout history, wars have contributed mightily to the knowledge needed for man-inventions. Remotes were developed to help the military explode bombs and reroute ships. After the fighting ceased, intrepid scientists tried to find better ways of applying this knowledge. Remotely exploding a bomb is a bit violent, but changing the channel on my TV is completely safe. Out of the ashes of warfare arose one of the ultimate cool boy's toys.

The most basic television remotes use infrared (IR) light. Infrared is a portion of the electromagnetic spectrum just outside of visible light. Inside the remote control you have electrical contacts that correspond to your touch pad. Pressing a button sends a signal to the microprocessor of the remote. The microprocessor then sends a binary code – a series of ones and zeroes – to the light-emitting diode (LED) on the end of the remote. Every command has a different binary signal that contains a code for the particular device and then a series of ones and zeroes that correspond to the desired action. It then finishes off with an end code for the device. The light essentially flashes on and off rapidly (and in the correct sequence). A receiver on the front of the television picks up the flashing light and the signal is relayed to the television's microprocessor. You can't see the flashing light because IR is invisible to our eye. Even if you could see it, the flashes are too fast for your eye to detect. You get to surf, change the volume and mute the TV, all at the push of a button.

IR remotes have a few downsides. One, they are basically line of sight. A stronger LED will allow the aim to be off a little, but the TV's eye still needs to see the signal. Second, interference from other IR sources has to be limited. You can accomplish this in a couple of ways. The remote uses only one frequency of light and the TV can only accept the same frequency. Also, watching TV indoors is a good idea, since the sun gives off IR waves and IR remotes are limited to about ten metres of effective range.

The IR remote is the old standard, but remote technology is getting better. I look forward to the day where we won't even need to push buttons; my thumb gets tired easily. We will just think about changing the channel and our television will start scanning the airwaves. When we reach that level, watching TV with your hyperactive best friend would be a total pain in the arse.

Brain fart

Infrared waves not only help your remote control, they also keep sandwiches warm for days on end at your favourite lunchtime cafe.

What about RF remote controls then?

Radio frequency (RF) remotes are the second-most-popular type. Key fobs, garage-door openers and remote-control toys all use RF remotes. RF remotes use a radio frequency to deliver the signal. After that, they work essentially the same as the IR remotes with one big advantage: range.

RF remotes have an effective range of about thirty yards. The signal will also go around corners and through walls. Since they aren't line of sight, you don't have to lift your arm at weird angles to get the device to work. Many high-end audiovisual components now come with RF remotes.

Garage-door remotes and key fobs are now sending a coded signal to deter thieves. In the old days, you could drive around the neighbourhood and probably open several other doors with your remote garage-door opener. The codes are scrambled now and only the correct transmitter can activate the opener.

Key fobs are also more user-friendly today. I drive one of the most manly cars on the planet: a white van. Do you know how many white vans are on the road? On any given Saturday in a shopping centre car park you might see a thousand. Luckily, one press of the button and my car lights flash and the doors open. I haven't lost my car yet and the door is always unlocked waiting for me. If we had bought the deluxe van, I wouldn't even have to open the door with my hand at all. Another press of the button and the door would magically open.

So what's the best remote in the universe?

Universal remote controls give you the ability to control everything in your life. For a few years they were the exclusive preserve of the mega-rich and the mega-lazy, but not any more. Gone are the days of a coffee table full of remote controls. It used to take my best friend six minutes to play a CD because of the twelve remotes that all had to be used in the correct sequence. It was probably easier to launch a nuclear missile than to hear music at his house.

New remotes also have learning capabilities. You just aim another remote at the new one and the new one will learn all of the codes for each desired feature. It will take a few days to program, but it beats having twelve different remotes. Universal remotes can be either IR or RF style remotes. The technology is the same; universal remotes just have more buttons.

Many universal remotes now come with LCD touchpads to make surfing easier. Touching the pad causes the circuit to be complete and the signal is sent to the correct toy (oops, electrical component). They are easier to see in the dark because the pad is lit and they are a great way to make your male friends jealous. Of course, first dates may think you are a complete saddo and never want to see you again, but dating another techno-geek is a great option. They might actually be impressed with your toys and almost always earn quite a good salary to boot.

Now if they could just put a beeper on the remote so you won't lose it. Even universal remotes are still going to slip down the back of the sofa. Maybe we need scientists to get to work on the brain-driven remote control. If you are a scientist, put down this book and go to work. Now!

Why is duct/duck tape so great?

Duct (or duck) tape is the tape of choice for all intrepid men. A single roll can clear a tiresome to-do list faster than anything. All men should have a roll in the house and one in the car. There is quite a bit of disagreement over the name and origin of duct tape, but either way, men are glad it was invented.

It is commonly thought that 'duck tape' was originally developed by the Johnson & Johnson Company during World War II to seal ammunition boxes. This makes sense, since duck tape is just a waterproof version of the medical tape J&J was already supplying to the US military. The dispute over the real name is a bit more complicated.

The name is thought to come from one of two sources. One, since it was waterproof, the water rolls off it like off a duck's back. Or two, the cloth part of the tape is made of cotton duck (a type of canvas). Either way, the GIs soon found out that it was good for a million different uses. The term duct tape shows up later when it was used to seal air-conditioning ducts. The name Duck Tape was trademarked in the 1980s by a company that specializes in the tape (not J&J).

What makes it so great? The tape is composed of three layers. The inner layer is a rubber-based adhesive. The middle is the fabric cloth that lends strength and allows the tape to tear easily. The outside is a waterproof vinyl now available in a multitude of colours, although actual duct tape is a bright silver colour. Waterproof, tough and easy to tear makes duck tape a winner in a man's world. There are tapes that do each trait better than duck tape, but not a single tape that is more versatile.

Duck tape has legions of adoring fans. Entire books have been written to pay homage to that amazing tape. Duck tape or duct tape – either way, men love it. I am pretty sure somewhere in the annals of history an actual duck was taped with duck tape, but I have not been able to find any credible research on the subject. It has to work; duck tape sticks to everything. If you see a silver-coated duck walking around town, you will know that I was just doing some research and not trying to stick it to the insides of a turkey. No actual ducks were harmed in the making of this book.

How does WD-40 work?

WD-40 has been a staple of households for years and the blue and yellow can has ended more annoying squeaks than anything in the history of civilization. One spray and the creaks are gone. Things just move easier under the influence of WD-40. In the states, Duck Tape and WD-40 have been called the 'redneck repair kit', but no matter where he is, every man needs a can and a roll to make his life complete.

Take some aliphatic hydrocarbons, petroleum oil, carbon dioxide and a few other ingredients. Mix together and you have WD-40. Aliphatic hydrocarbons are long, straight-chained hydrocarbons. These hydrocarbons are flammable (as any WD-40 fan will tell you). These hydrocarbons are in the same family as wax, so they are a successful lubricant. Carbon dioxide is the primary propellant used to launch that great liquid out of the can. Petroleum oil is pretty much the same as the good old-fashioned penetrating oil that your grandpa used. Of course, after 1958 he probably used WD-40.

WD-40 stands for Water Displacement perfected on the fortieth try. I for one am glad that it was finished on the fortieth try; WD-38 just doesn't roll off the tongue. Originally designed by the Rocket Chemical Company – which actually changed their name to the WD-40 Company – it was designed as a rust-prevention solvent for the aerospace industry and was used on the Atlas ICBM to stop corrosion and rust. It worked for that but also found a million other uses, including fixing squeaky hinges, removing tar and loosening nuts. Many of the engineers took cans home to fix all manner of problems. In 1958, the company started selling it to the public and a legend was born. The secret formula is still in use today.

A few brand names have been added to the WD-40 Company brand family over the years, but WD-40 will always be near and dear to the male heart.

My favourite use of the magic formula is as fuel for a spud gun, but many uses are safer. Removing crayon marks from all manners of stuff is a great use for a new dad. I also like spraying the top of my birdfeeder, so squirrels just slide off. For you gamblers out there, it also cleans dice and pool cues. However, there is no proof that is a cure-all for arthritis or glue-ear.

That familiar blue and yellow can is a staple of manhood. If you want something to move, just grab that can. A can also makes the perfect gift idea for any mechanic or general DIY freak. My dad is hard to buy for, so last year I got him a can of WD-40 and a roll of duct tape. It was the best present he opened and two gifts he is sure to use. He can use them to maintain the new vacuum cleaner I bought my mum.

Did you know?

WD-40 has been used to remove a boa constrictor from a car engine, lubricate inner tubes and clean prosthetic limbs!

The clash of the titans: what happens if you spray duct tape with WD-40?

The space-time continuum is based on the assumption that everything can be described by four dimensions. Length, width and height will give you the where, and time, the fourth dimension, will give you the when. You can only be in any one place at one time in your life. We use the space-time continuum every time we make plans with our friends. We agree on a time and a physical location – four dimensions! If you get the day wrong, you may be by yourself.

In our everyday life, time is constant, but as an object approaches the speed of light, time, length and even mass change. Wow! Time takes on a new meaning since it is related to velocity. Einstein garnered most of his fame dealing with this thought, although the crazy hair helped his celebrity. If he were bald, he probably wouldn't have been famous. Experiments have proved his theory to be correct. The space-time continuum also shows up in virtually every sci-fi book or movie. The continuum lends itself so well to time travel. Ten thousand years ago, where you sit now might have been underwater. In books and movies, you can travel back in time and make small changes that impact loads of future decisions. We love it and we go along for the ride.

WD-40 and duct tape are complete opposites. No mixing of these two is allowed. In a simple world, spraying WD-40 onto duct tape would just cancel both out. Think of it as eating ice cream while jogging on a treadmill, but in man land it is more complicated.

Unsurprisingly, I have my own private, non-scientific theory. Mixing the two is like seeing yourself during time travel – guaranteed to destroy the universe. The initial spray onto the duct tape will appear normal, but rapidly you will rip holes in the space-time continuum. The spray would create a mini black hole. Time slows down as it nears the warped space created by a black hole. The black hole would slowly pull nearby objects in. As you are pulled into the black hole, you would be spaghettified as the intense gravity stretches you out.

Please, for the sake of humanity, never spray duct tape with WD-40. Our world may have issues, but we don't want to destroy it. I recommend storing duct tape and WD-40 in separate locations in your house to minimize the chances of an accidental rip in the continuum. Of course, maybe spraying WD-40 on duct tape won't do anything, but is it really worth taking the chance?

Scientifically speaking

Space isn't remote at all. It's only an hour's drive away if your car could go straight upwards.
- Sir Fred Hoyle

Why is war good for science and technology?

Speaking of the battle between duct tape and WD-40, wars do more to help technology than most peacetime research has ever done. This is sad but true. During times of war, we dedicate vast sums of money and brainpower into focused pursuits. Of course, most of these pursuits are designed to give us an advantage over the enemy, which is a politically correct term for better ways to kill people.

Nuclear power, microwave ovens, remote controls, duct tape and WD-40 are just a few things either invented in times of war or devices that grew out of military-industrial research. WD-40 grew out of the Cold War, a war that involved no killing but the deployment of tons of new technology. The arms and space races were a boon to blokes. We got tons of new gizmos with very little loss of life, a win-win situation. Most problems in the world can be solved if we focus our money and brains on them.

Many of these same toys may also be our downfall. Is the next Isaac Newton spending twenty-two hours a day slouched in a semi-coma trying to conquer the next computer game craze? Is the next James Dyson too busy to invent because he is watching *Spider-Man 7* in his home cinema for the ninety-second time that month?

On the other hand, perhaps playing with boys' toys will lead us to new ways to improve the world. Hopefully we will never stop inventing, discovering and playing. Out of this playtime will come new things to amuse us. And a whole load more pub science to explain it all.

Scientifically speaking

Innocence about science is the worst crime today

- Sir Charles Percy Snow

WEB RECOURCES

About.com: Chemistry: www.chemistry.about.com
American Dental Association: www.ada.org
Brainy Quote: www.brainyquote.com
ConsumerAffairs.com: www.consumeraffairs.com
eHow: www.ehow.co.uk
Helium: www.helium.com
Inquiry Journal: www.unh.edu/inquiryjournal
Inventors Digest www.inventorsdigest.com
Live Science: www.livescience.com
National Hot Rod Association: www.nhra.com
Science Museum of Minnesota: www.smm.org
The Kitchen Project: www.kitchenproject.com
United States Golf Association: www.usga.org
WD-40 Company: www.wd40.com/uses
Wikipedia: www.wikipedia.org
Wise Geek: www.wisegeek.com

About the author

Bobby Mercer is an award-winning teacher, coach, author and dad. He is the author of *Quarterback Dad: A Play-by-Play Guide to Tackling Your New Baby*, a fun look at fatherhood using American football terms. He has a science education degree from UCF and a master's degree in physics education from the University of Virginia. He is also the author of two juvenile science books: *Smash It! Crash It! Launch It!: 50 Mind-Blowing Eye-Popping Science Experiments* and *The Leaping, Sliding, Sprinting, Jumping Science Book: 50 Super Sports Science Activities*
He lives outside of Asheville, North Carolina, with his family (www .bobbymercerbooks.com).

INDEX